Grade 1

Reveal MATH®

Differentiation Resource Book

McGraw Hill

mheducation.com/prek-12

McGraw Hill

Send all inquiries to:
McGraw-Hill Education
8787 Orion Place
Columbus, OH 43240

ISBN: 978-1-26-421061-9
MHID: 1-26-421061-2

Printed in the United States of America.

1 2 3 4 5 6 7 8 9 LHN 25 24 23 22 21 20

Grade 1

Table of Contents

Unit **4**

Addition within 20: Facts and Strategies

Lessons

Unit **5**

Subtraction within 20: Facts and Strategies

Lessons

Unit 6
Shapes and Solids

Lessons

Unit 7
Meanings of Addition

Lessons

Unit 8
Meanings of Subtraction

Lessons

Unit 9
Addition within 100

Lessons

Unit 10
Compare Using Addition and Subtraction

Lessons

Unit **11**

Subtraction within 100

Lessons

Unit **12**

Measurement and Data

Lessons

Unit **13**

Equal Shares

Lessons

Counting Patterns to 100

Name ______________________

Review

You can use a number chart to count.

21	22	23	24	25	26	27	28	29	30
31	32	33	34	35	36	37	38	39	40
41	42	43	44	45	46	47	48	49	50

The tens stay the same. The ones go up by 1.

The ones stay the same. The tens go up by 1.

Count. Use the counting pattern to complete. Circle the number that comes next.

1. 15, 16, 17, 18, 19,

1	2	3	4	5	6	7	8	9	10
11	12	13	14	15	16	17	18	19	20
21	22	23	24	25	26	27	28	29	30

2. 87, 88, 89, 90, 91,

81	82	83	84	85	86	87	88	89	90
91	92	93	94	95	96	97	98	99	100

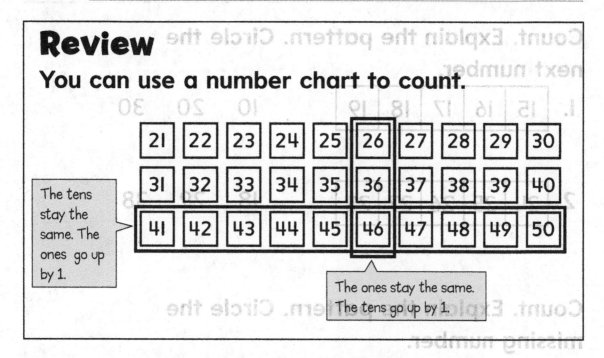

Counting Patterns to 100

Name ______________________________

Count. Explain the pattern. Circle the next number.

1.	15	16	17	18	19	10	20	30
2.	24	25	26	27	28	18	29	38

Count. Explain the pattern. Circle the missing number.

3.	38	?	40	41	42	39	43	49
4.	95	96	?	98	99	84	87	97
5.	66	67	68	69	?	60	70	79

Lesson **2-2 • Reinforce Understanding**

Patterns on a Number Chart

Name ______________________________

Review

Kyle collects marbles.
There are 108 marbles in a jar.

He starts at 108.

What are the next 4 numbers he counts?

1	2	3	4	5	6	7	8	9	10
11	12	13	14	15	16	17	18	19	20
21	22	23	24	25	26	27	28	29	30
31	32	33	34	35	36	37	38	39	40
41	42	43	44	45	46	47	48	49	50
51	52	53	54	55	56	57	58	59	60
61	62	63	64	65	66	67	68	69	70
71	72	73	74	75	76	77	78	79	80
81	82	83	84	85	86	87	88	89	90
91	92	93	94	95	96	97	98	99	100
101	102	103	104	105	106	107	108	109	110
111	112	113	114	115	116	117	118	119	120

The tens stay the same until the ones start back at 0. Then the tens go up by 1.

The next 4 numbers are 109, 110, 111, 112.

Lin counts her marbles.

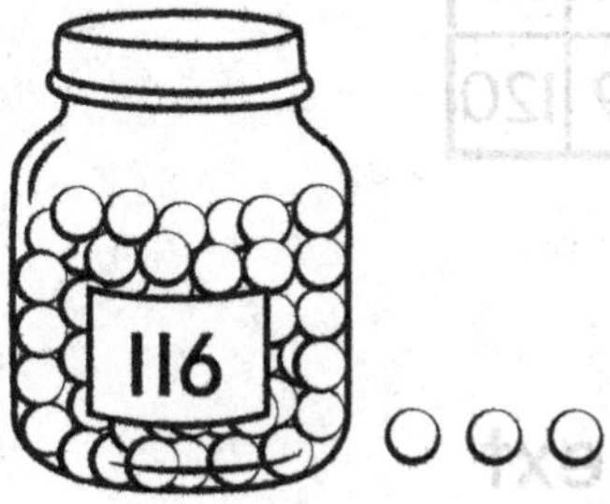

1. What number does Lin start counting from?

A. 3 B. 116 C. 119

2. What are the next 3 numbers she counts?

A. 4, 5, 6 B. 116, 117, 118 C. 117, 118, 119

Patterns on a Number Chart to 120

Name ______________________________

Suman is counting with his mom. His mom asks him to start counting from different numbers.

1	2	3	4	5	6	7	8	9	10
11	12	13	14	15	16	17	18	19	20
21	22	23	24	25	26	27	28	29	30
31	32	33	34	35	36	37	38	39	40
41	42	43	44	45	46	47	48	49	50
51	52	53	54	55	56	57	58	59	60
61	62	63	64	65	66	67	68	69	70
71	72	73	74	75	76	77	78	79	80
81	82	83	84	85	86	87	88	89	90
91	92	93	94	95	96	97	98	99	100
101	102	103	104	105	106	107	108	109	110
111	112	113	114	115	116	117	118	119	120

Count and color numbers.

1. Start at 19. Count and color the next 4 numbers yellow.

2. Start at 33. Count and color the next 6 numbers blue.

3. Start at 68. Count and color the next 5 numbers red.

Patterns on a Number Line

Name ______________________________

Review

Number lines can show counting patterns.

60 61 62 63 64 65 66 67 68 69 70 71

The tens digit stays the same. The ones digit goes up by 1.

Then the tens digit goes up by 1. The ones digit starts back at 0.

Complete the sentence.

1. 28 2☐ 30 31 32 33 34 35 36 37 38 39

The missing ones digit is ______.

2. 83 84 85 86 87 88 89 ☐0 91 92 93

The missing tens digit is ______.

3. What is the missing number?

106 107 108 109 ☐ 111 112 113 114 115 116

A. 105 B. 110 C. 117

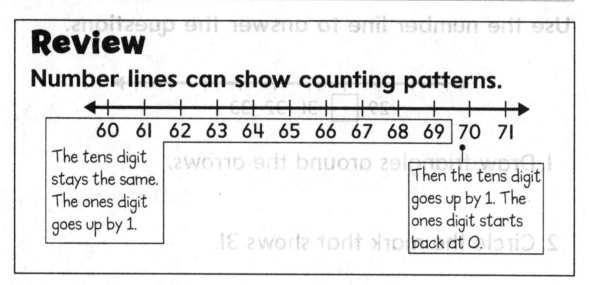

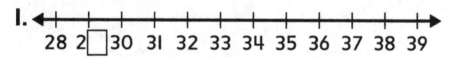

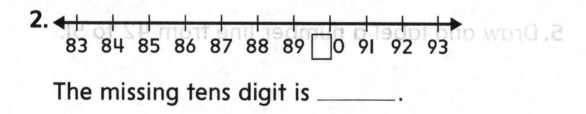

Patterns on a Number Line

Name _____

Use the number line to answer the questions.

29 ☐ 31 32 33

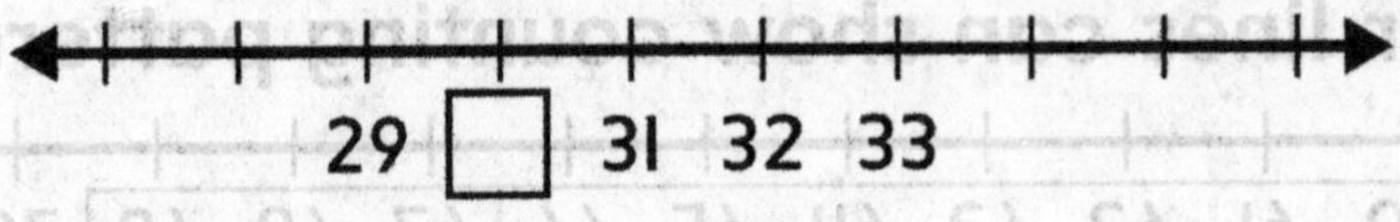

1. Draw triangles around the arrows.

2. Circle the mark that shows 31.

3. Color the space between 32 and 33.

4. Fill in the number that goes in the box. Explain.

5. Draw and label a number line from 42 to 51.

Patterns When Reading and Writing Numbers

Name ______________________________

Review

You can use counting patterns to help you read and write numbers.

61	62	63	64	65	66	67	68	69	70
71	72	73	74	75	76	77	78	79	80
81	82	83	84	85	86	87	88	89	90
91	92	93	94	95	96	97	98	99	100

The ones go up by 1 from 1 to 9, then start over at 0.

The tens go up by 1 when the ones start over at 0.

Write the next 3 numbers.

1. | 54 | 55 | 56 | ______, ______, ______

2. | 115 | 116 | 117 | ______, ______, ______

3. | 77 | 78 | 79 | ______, ______, ______

4. | 8 | 9 | 10 | ______, ______, ______

5. | 35 | 36 | 37 | ______, ______, ______

Patterns When Reading and Writing Numbers

Name ______________________

Deondre helps out at the library. He returns books to the shelves.

Draw books 87, 46, 113, and 93 where they belong.

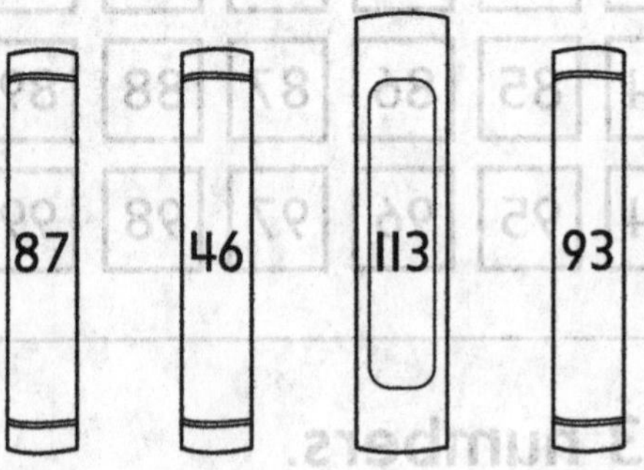

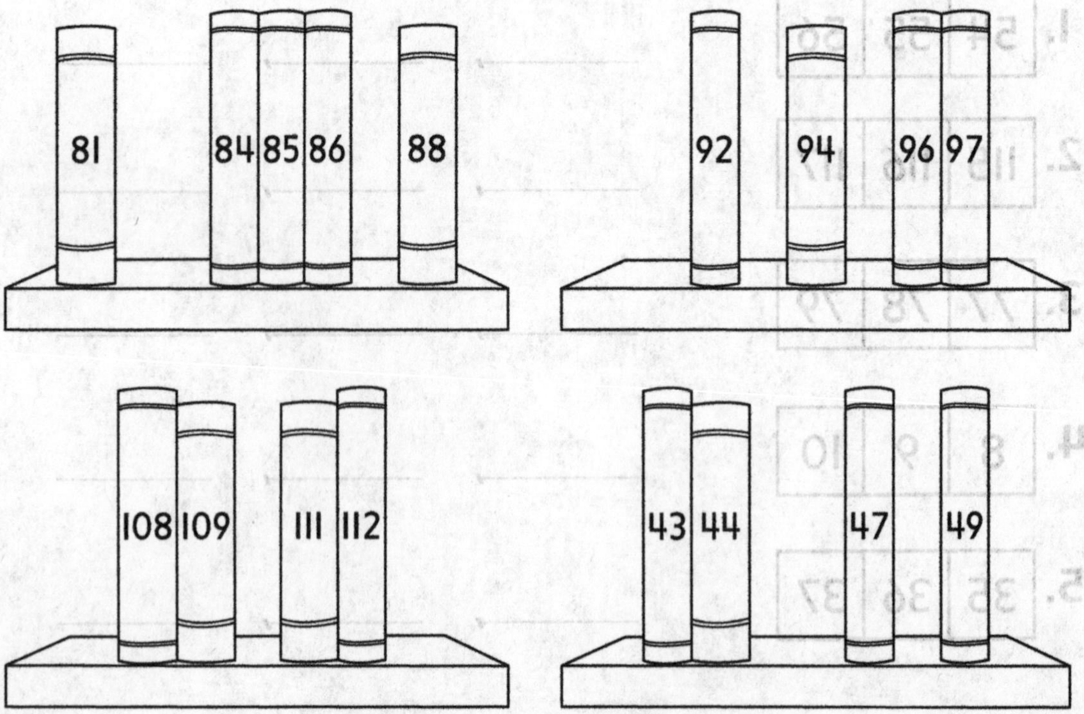

Patterns When Representing Objects in a Group

Name ______________________________

Review

Putting objects in groups can help you count.

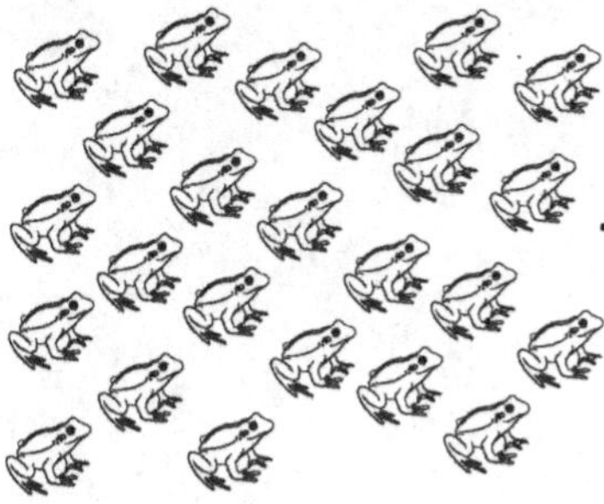

→

										10
										20
				25	26	27	28	29	30	24
31	32	33	34	35	36	37	38	39	40	
41	42	43	44	45	46	47	48	49	50	

Count the objects. Color each number on the chart as you count. Write a number to show how many.

1\.  ______

1	2	3	4	5	6	7	8	9	10
11	12	13	14	15	16	17	18	19	20
21	22	23	24	25	26	27	28	29	30

Patterns When Representing Objects in a Group

Name ____________________

Armand is making fun packs for campers. He groups items in 10s. Circle groups of 10s for the objects.

1. 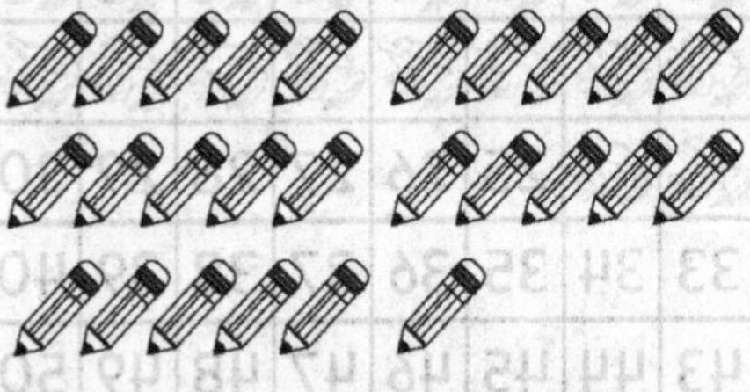

2. 

3. 

4. Armand puts 2 toy cars, 1 pencil, and 1 apple in each pack. How many fun packs can he make?

_______ fun packs

Numbers 11 to 19

Name ______________________________

Review

You can show 10 ones as 1 group of ten.

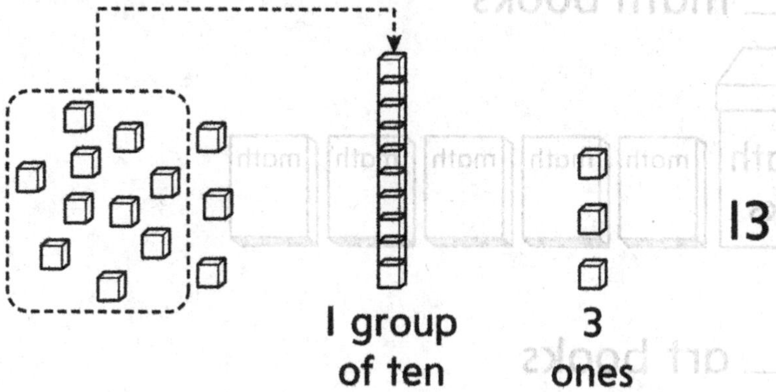

Teen numbers have 1 group of ten and some ones.

How many groups of ten and how many ones? Write the numbers.

1. 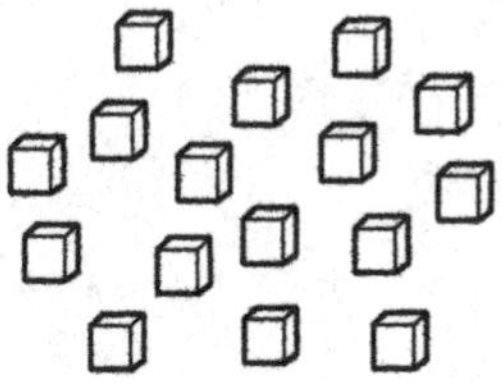

   ______ group of ten and ______ ones is ______.

2. 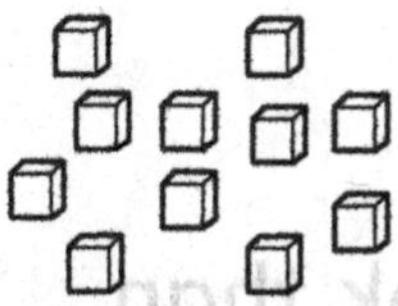

   ______ group of ten and ______ one is ______.

3. Draw blocks to show 17. Write the number.

______ group of ten and ______ ones is ______.

Numbers 11 to 19

Name ______________________________

The tutor keeps sets of 10 books in a box.
Fill in the number of his books.

1. ______ math books

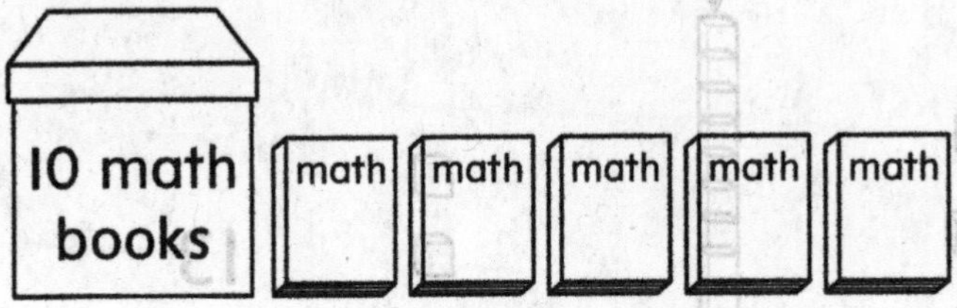

2. ______ art books

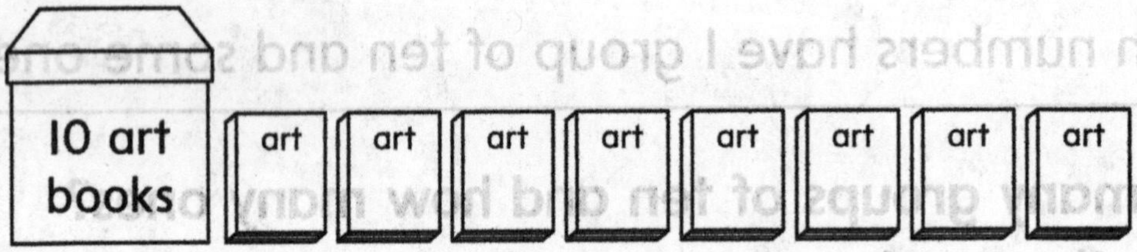

3. ______ science books

4. The tutor has one more reading book than science books. Does he have more reading books or math books? Explain you thinking.

Understand Tens

Name ______________________

Review

10 ones make 1 ten.

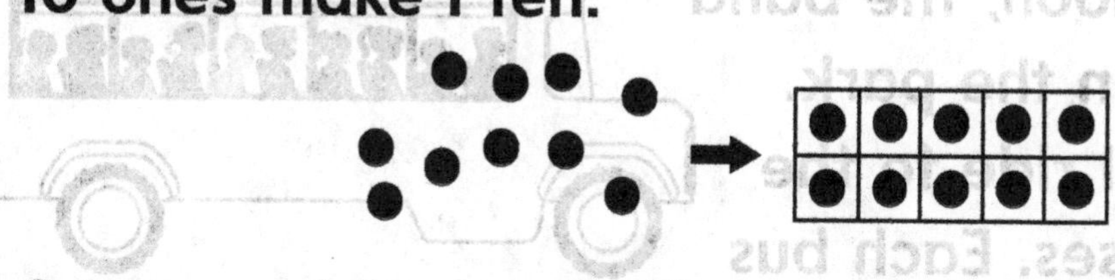

2 tens and 0 ones is 20.

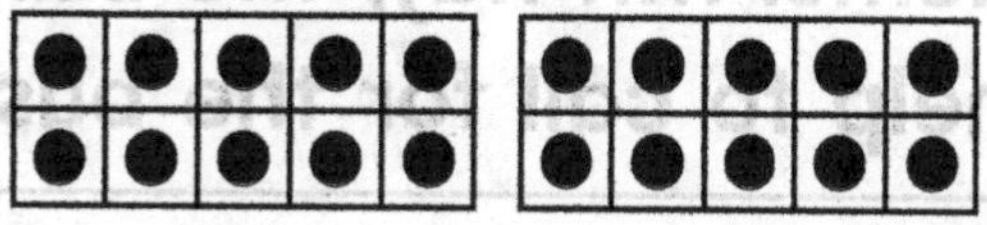

What is the number?

1. 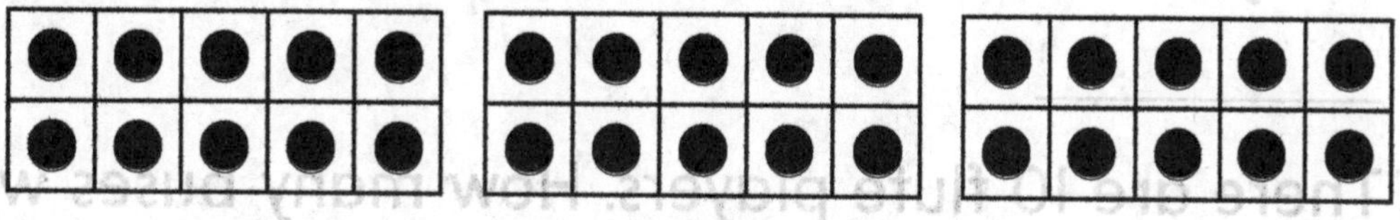

____ tens and ____ ones is ____.

2. 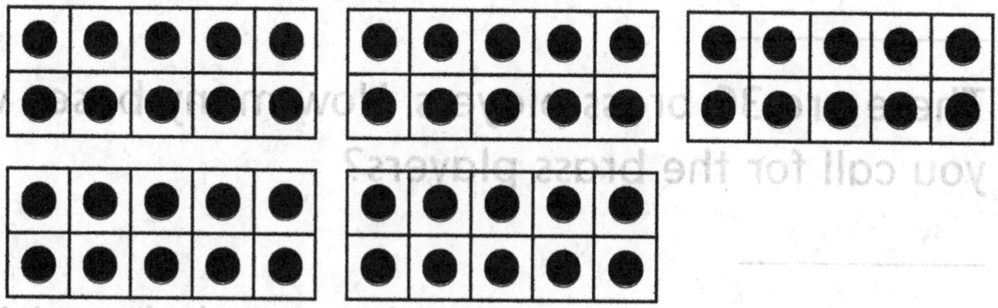

____ tens and ____ ones is ____.

3. 7 tens and 0 ones is ____.

4. 9 tens and 0 ones is ____.

5. 4 tens and 0 ones is ____.

Understand Tens

Name ______________________________

Review

This afternoon, the band is playing in the park. Players will ride to the park in buses. Each bus holds 10 students. Mr. Ray, the band director, needs your help to call for the buses.

1. There are 20 clarinet players. How many buses will you call for the clarinet players?

   ____

2. There are 10 flute players. How many buses will you call for the flute players?

   ____

3. There are 30 brass players. How many buses will you call for the brass players?

   ____

4. At the park, Mr. Ray will ask you to help put the players equally on both sides of the stage. How could they sit? Explain your thinking.

Represent Tens and Ones

Name ____________________

Review

45 is 4 tens and 5 ones.

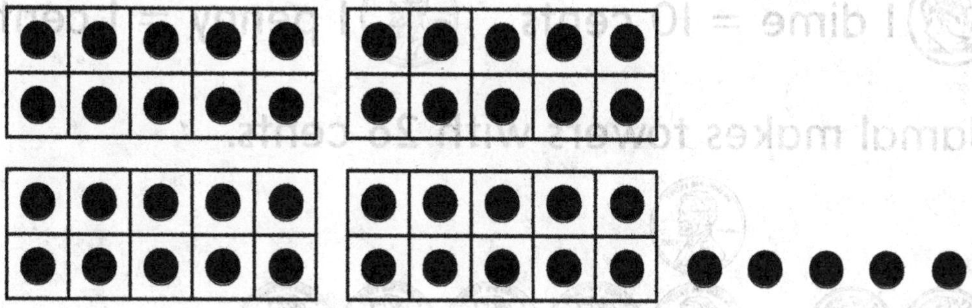

How many? Write the numbers.

1.

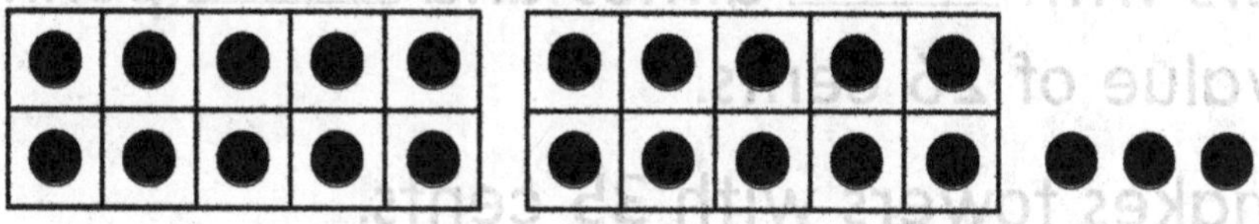

_____ tens and _____ ones is _____.

2.

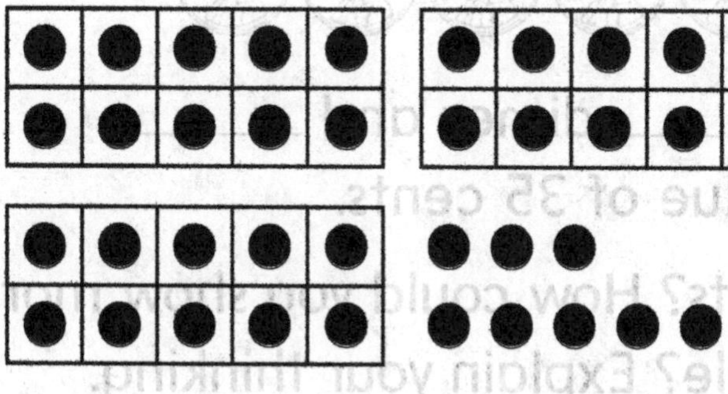

_____ tens and _____ ones is _____.

3.

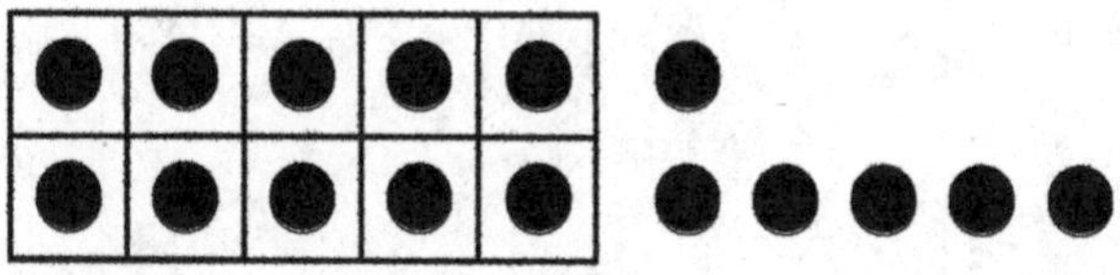

_____ ten and _____ ones is _____.

Represent Tens and Ones

Name ______________________________

Jamal and Nicole are making towers of coins. Use the values to find out who has more cents.

 I dime = 10 cents  I penny = I cent

I. Jamal makes towers with 26 cents.

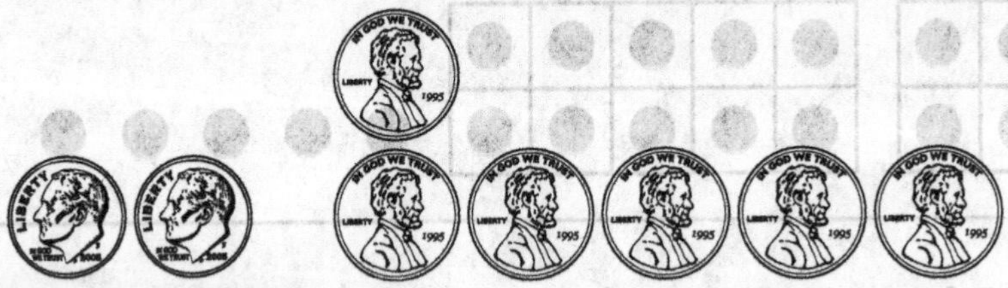

His towers with ______ dimes and ______ pennies have a value of 26 cents.

2. Nicole makes towers with 35 cents.

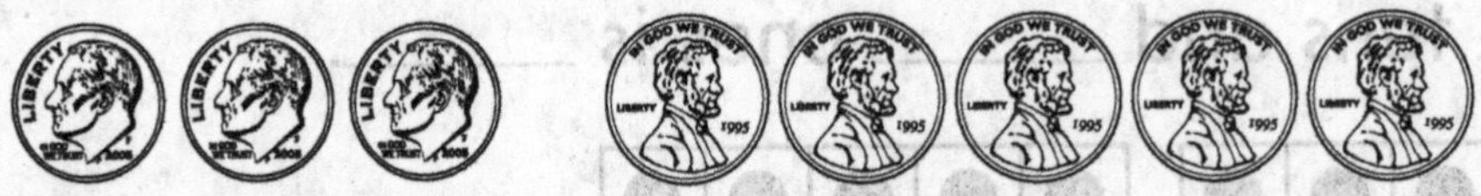

Her towers with ______ dimes and ______ pennies have a value of 35 cents.

3. Who has more cents? How could you show more than Jamal or Nicole? Explain your thinking.

Represent 2-Digit Numbers

Name ______________________________

Review

2 tens and 5 ones is 25.

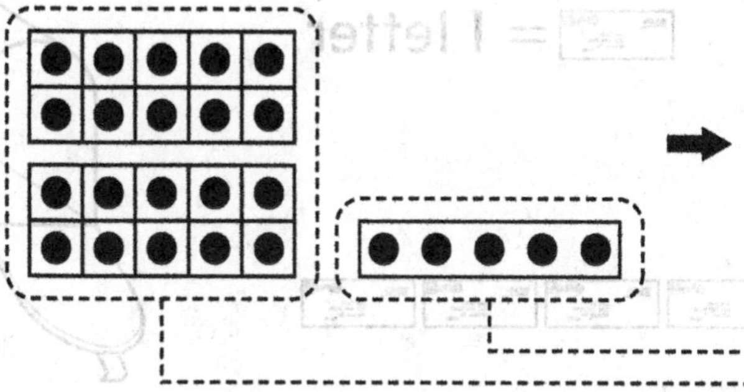

tens	ones
2	5

How many tens? How many ones?

1.

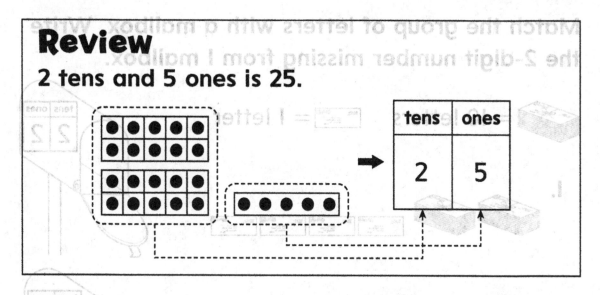

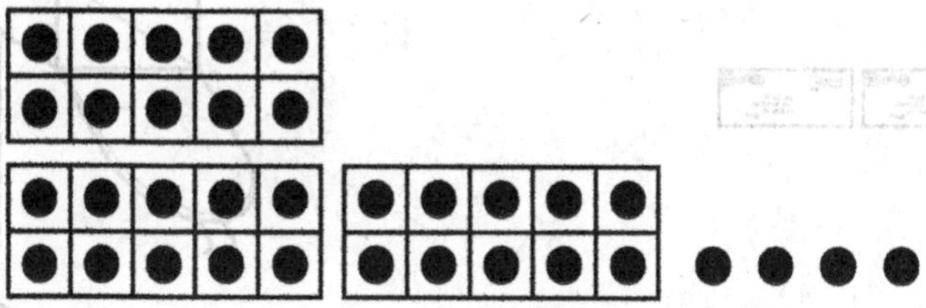

tens	ones
___	___

2. 97

tens	ones
___	___

3. 52

tens	ones
___	___

Represent 2-Digit Numbers

Name ______________________

Match the group of letters with a mailbox. Write the 2-digit number missing from 1 mailbox.

 = 10 letters  = 1 letter

1. 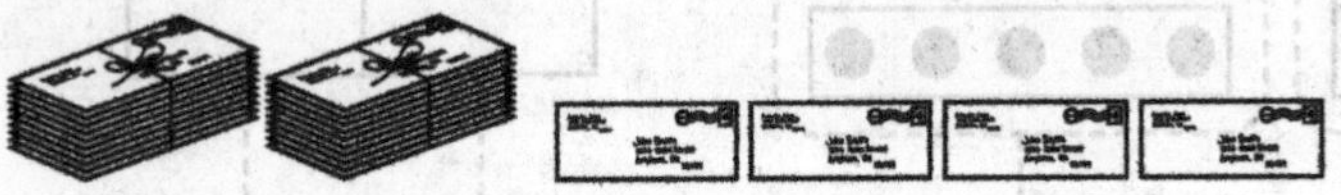

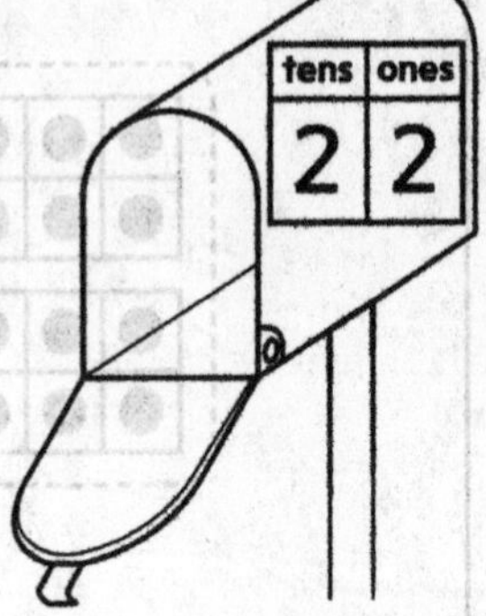

2. 

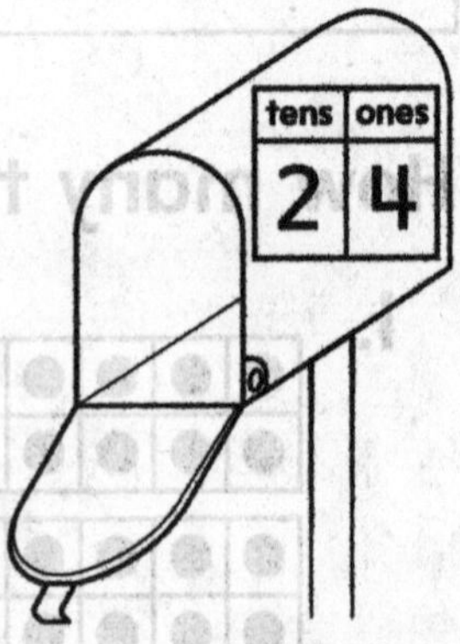

3. 

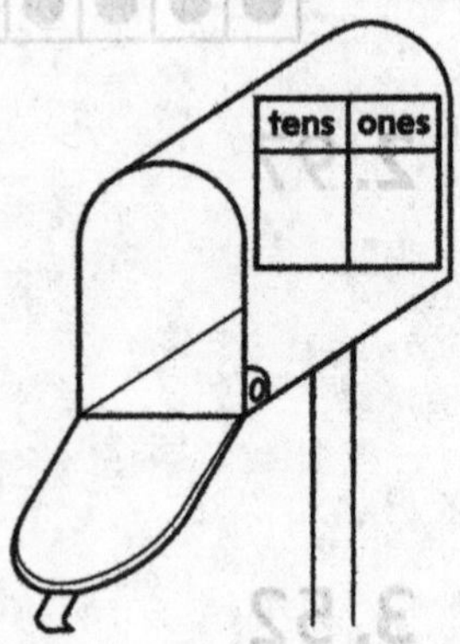

4. Which mailbox has the fewest letters? Explain.

Represent 2-Digit Numbers in Different Ways

Name ______________________________

Review

You can show the same number in different ways.

23 is 2 tens and 3 ones

23 is also 1 ten and 13 ones

Circle all the ways to show the number.

1. 41

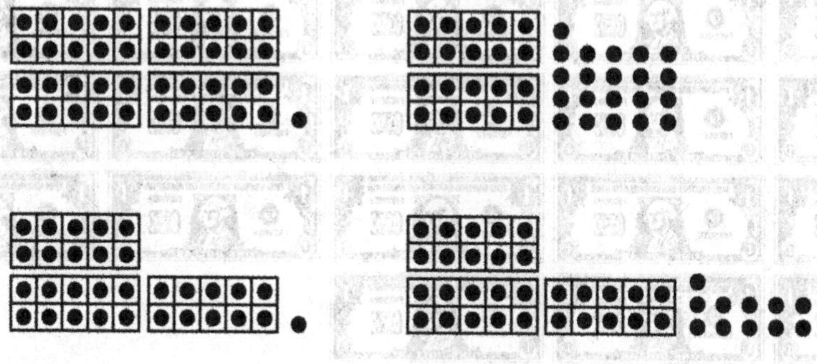

2. 36

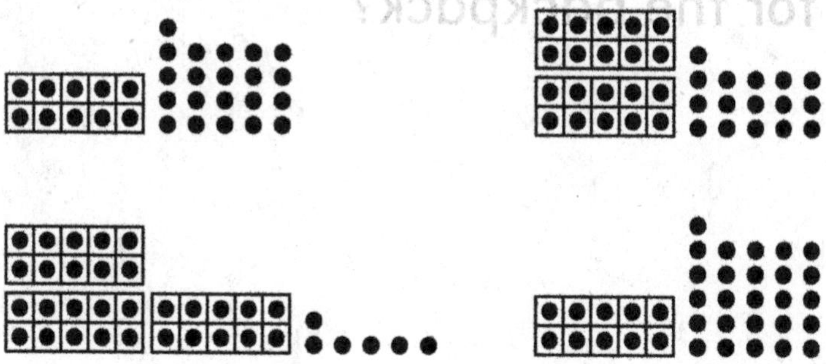

Lesson 3-5 • Extend Thinking

Represent 2-Digit Numbers in Different Ways

Name ______________________________

Last week, Alex ripped his backpack. He needs a new one for school. The backpack he wants costs $32.

1. Choose all the ways Alex can pay exactly $32 for the backpack.

A. 

B. 

C. 

2. How did you decide the ways Alex can pay exactly $32 for the backpack?

Compare Numbers

Name ______________________

Review

You can compare two-digit numbers.

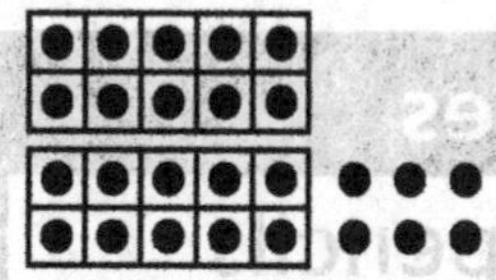

28
2 tens and 8 ones

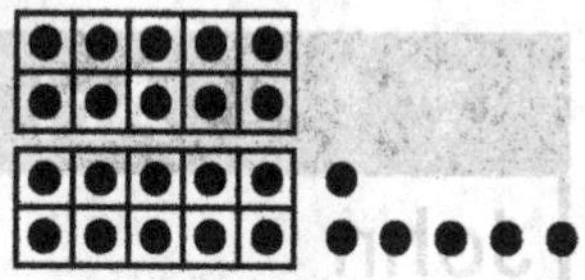

26
2 tens and 6 ones

Compare the **tens in 28** and **26.** Both have 2 tens.

Compare the **ones** in **28** and **26.**
8 ones are greater than 6 ones.

So, 28 is greater than 26. 26 is less than 28.

Write *greater than, less than,* or *equal to.*

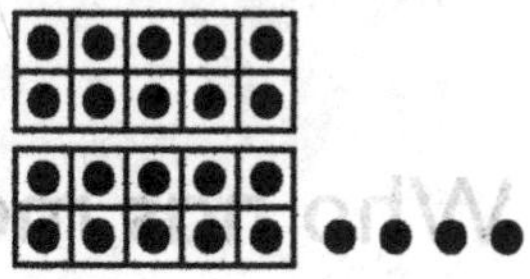

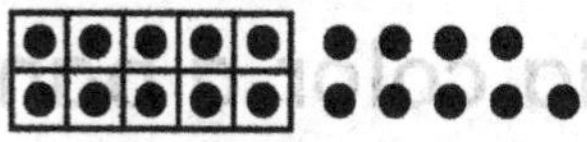

1. 24 is ______________________ 19
2. 81 is ______________________ 78
3. 59 is ______________________ 58
4. 32 is ______________________ 36
5. 85 is ______________________ 87

Compare Numbers

Name ____________________

The first graders are getting ready for an art show. How can you use the information in the table to find out who has more art supplies?

Art Supplies	
John	34 pencils
Liz	29 pencils
Nadia	50 crayons
Oni	43 markers
Sarah	54 crayons
Zack	41 markers

1. John and Liz draw animals. Who has more pencils? How do you know?

2. Sarah and Nadia color a banner. Who has more crayons? How do you know?

3. Zack and Oni make signs. Who has more markers? How do you know?

Compare Numbers on a Number Line

Name ____________________

Review

You can use a number line to compare numbers.
Compare 62 and 68.

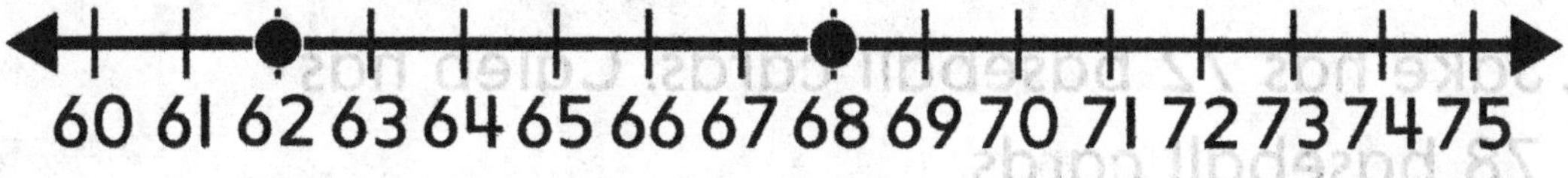

Greater numbers are to the right. The number 68 is to the right of 62. So, 68 is greater than 62.

Lesser numbers are to the left. The number 62 is to the left of 68. So, 62 is less than 68.

Use the number line to compare the numbers. Write *is greater than, is less than,* or *is equal to.*

1. 92 ____________________ 87.

2. 86 ____________________ 89.

3. 90 ____________________ 88.

Compare Numbers on a Number Line

Name ______________________________

Mr. Abi's students are talking about things they like to collect. How can you draw a number line with dots to show the collections? Compare the collections using *is greater than* and *is less than*.

1. Jake has 72 baseball cards. Caleb has 78 baseball cards.

2. Darcy has 43 action figures. Rio has 39 action figures.

Use Symbols to Compare Numbers

Name ______________________________

Review

You can use symbols to compare numbers.

6 tens is greater than 5 tens.

$62 > 51$

5 tens is less than 6 tens.

$51 < 62$

tens	ones
6	2

tens	ones
5	1

4 tens equals 4 tens.

3 ones equals 3 ones.

$43 = 43$

tens	ones
4	3

tens	ones
4	3

Compare the numbers. Write >, <, or =.

1.

tens	ones
3	8

◯

tens	ones
3	9

2.

tens	ones
9	4

◯

tens	ones
9	2

3. 87 ◯ 89

4. 45 ◯ 45

Use Symbols to Compare Numbers

Name ____________________

Taj is entering scoreboard data. How can you use >, <, or = to help him show each statistic?

1. At halftime, Galen has 11 points and Eve has 14 points.

   At halftime, Galen's points _____ Eve's points.

2. After period 3, Galen has 14 points and Eve still has 14 points.

   After period 3, Galen's points _____ Eve's points.

3. By the end of the game, the Lions score 30 points with 3-point shots. The Tigers score 36 points with 3-point shots.

   The number of Tiger 3-point shots _____ the number of Lion 3-point shots.

4. At the end of the game, Taj flashes up the score. Help him enter >, <, or =.

   LIONS 75 _____ TIGERS 57!

   LIONS WIN! LIONS WIN!

5. Help Taj write another statistic about the game using >, or <.

Relate Counting to Addition

Name ______________________________

Review

A number line can help you count and add.

How many balls?

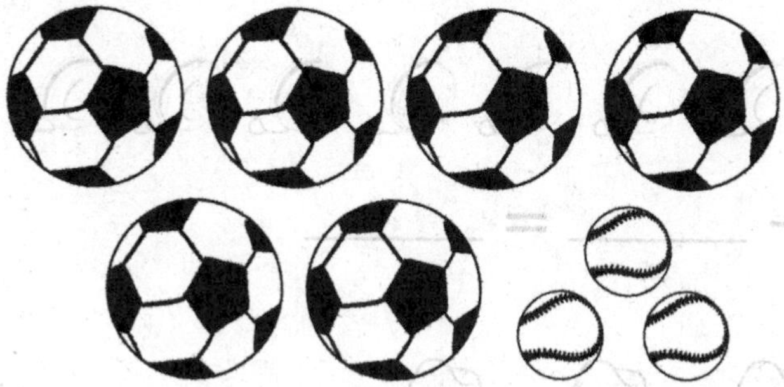

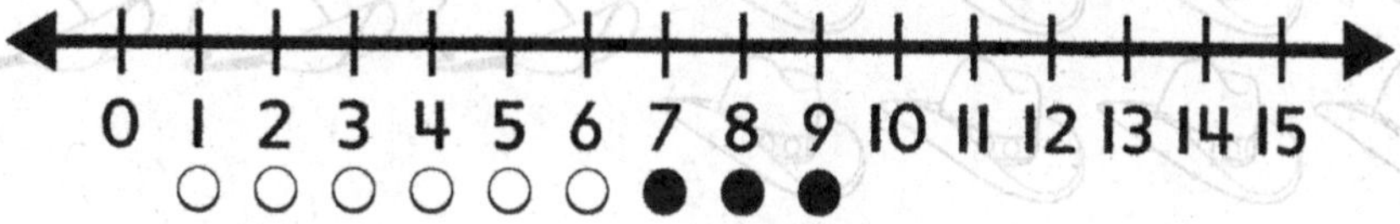

6 + 3 = 9

How many?

1.

______ books

2.

______ pieces of fruit

Relate Counting to Addition

Name ________________________________

Jay's mom buys party items. How many party items does she buy? Write numbers to match the picture.

1.

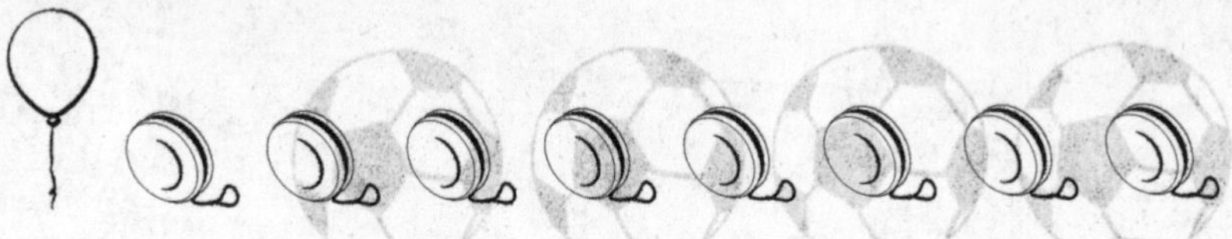

_____ + _____ = _____

2.

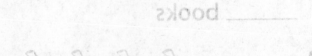

_____ + _____ = _____

3. The party has 8 children and 4 adults.

Show and count the people at the party.

Write an equation to show how many.

Count On to Add

Name ______________________________

Review

You can count on more quickly from the greater addend to find a sum.

You can find the sum of 3 + 5. Start with 5.
Count on 3 more to find the sum.

3 + 5 = 8

What is the sum? Use counters or the number line to count on.

1. 5 + 4 = ______

2. 4 + 7 = ______

3. 2 + 6 = ______

4. 9 + 1 = ______

Count On to Add

Name ______________________________

Kira is giving food to some of the zoo animals. How can you count on to add? Explain your thinking. Use the number line if you need help.

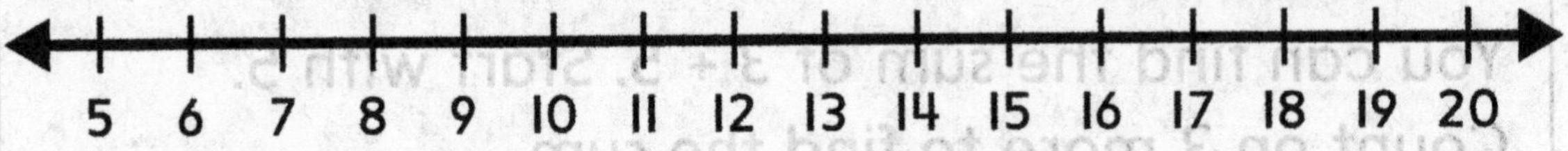

1. Kira feeds the monkeys 4 bananas.

   She feeds the gorillas 11 bananas.

   How many bananas does she feed them?

2. Kira feeds the elephants 14 bunches of carrots.

   She also feeds them 6 bunches of celery.

   How many bunches does she feed them?

3. Kira feeds the monkeys the same number of bananas again. How many bananas does she feed the animals?

Doubles

Name ____________________

Review

When the addends are the same, you can use doubles to add.

How many connecting cubes are there?

Both sets have 4 cubes.

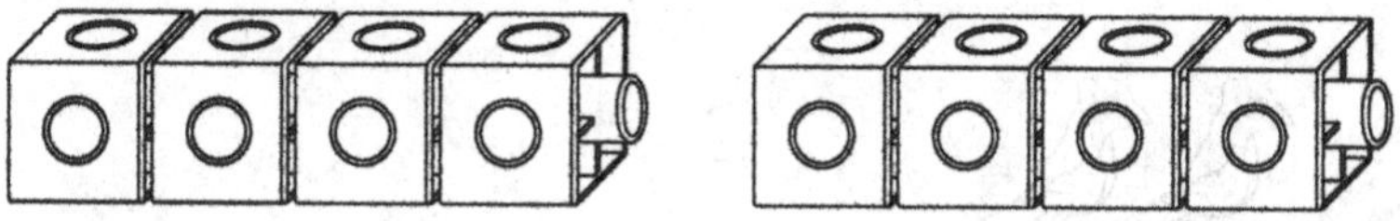

4 + 4 = 8

How can you use doubles to add?

1. 

   5 + 5 = ____

2. 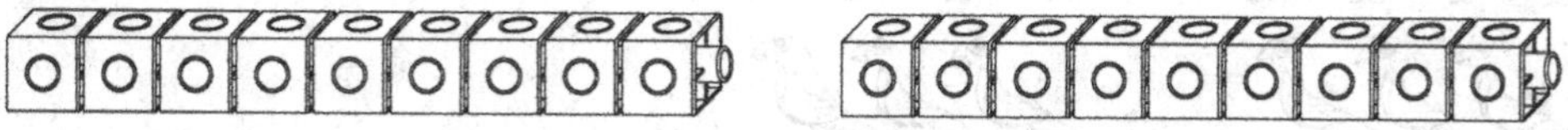

   ____ + 9 = ____

3. 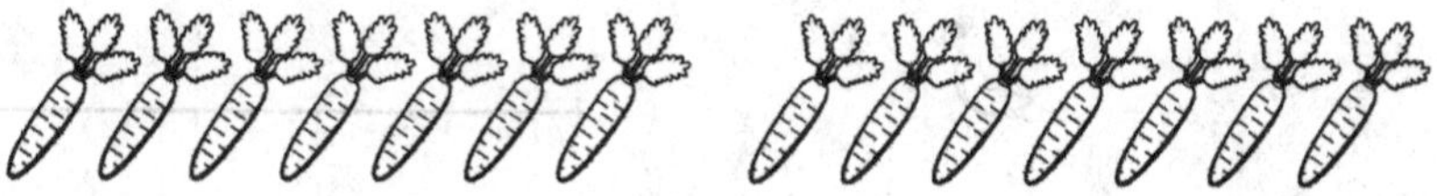

   ____ + ____ = ____

4. 10 + ____ = ____

5. ____ + 6 = ____

Doubles

Name ______________________

A group of birds is a *flock*. Match the flock of birds to an equation. How can you use doubles to find the sum? Write the missing equation.

1. 

$10 + 10 =$ ______

2. 

$5 + 5 =$ ______

3. 

____

4. 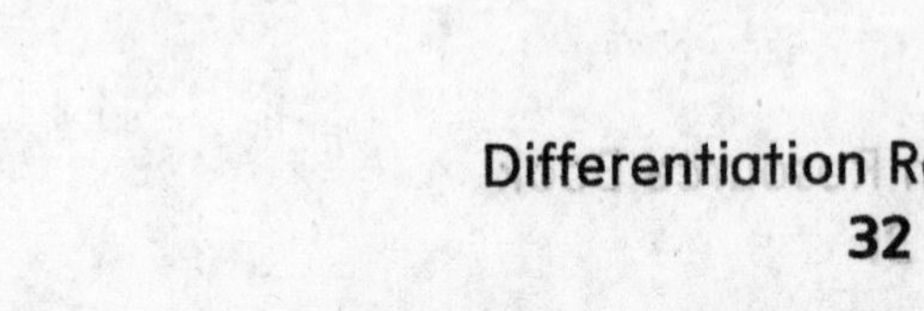

$6 + 6 =$ ______

Near Doubles

Name ______________________________

Review

You can use doubles to help you add.

10 + 8 = ?

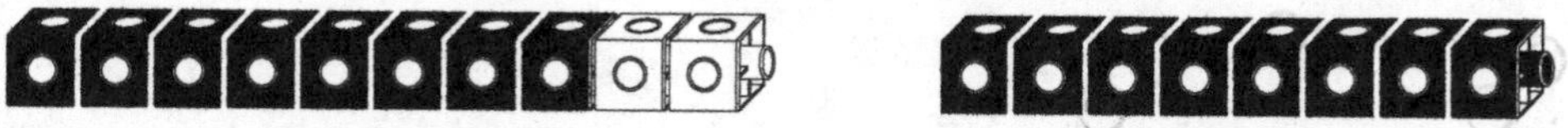

8 + 2 + 8

You know 8 + 8 = 16. 2 more is **18**.

How can you use doubles to add? Write numbers to match the problem.

1. 

   4 + _____ = _____. 1 more is _____.

2. 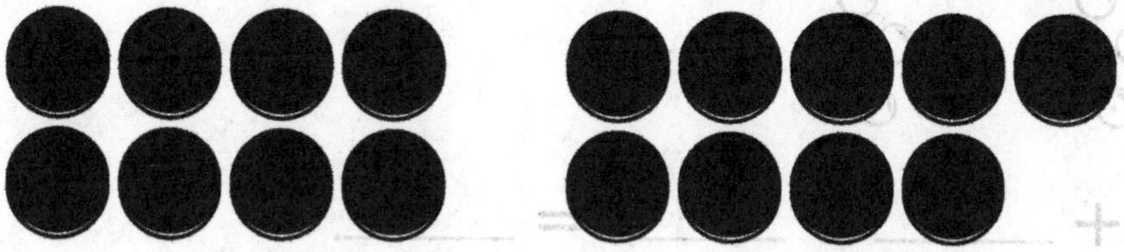

   _____ + 8 = _____. _____ more is _____.

3. 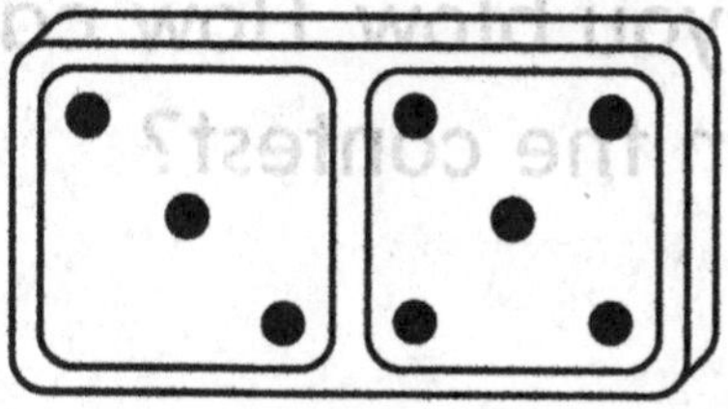

   _____ + 3 = _____. 2 more is _____.

Near Doubles

Name ______________________________

Some students have a contest to blow bubbles. How many bubbles do they blow?

1.

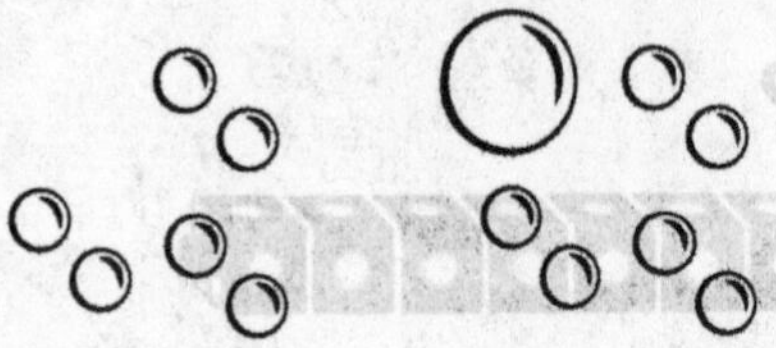

_____ + _____ + 1 = _____

2.

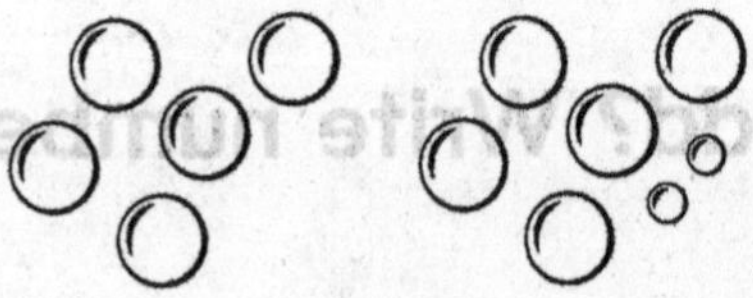

_____ + 5 + 2 = _____

3.

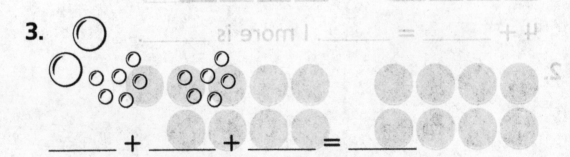

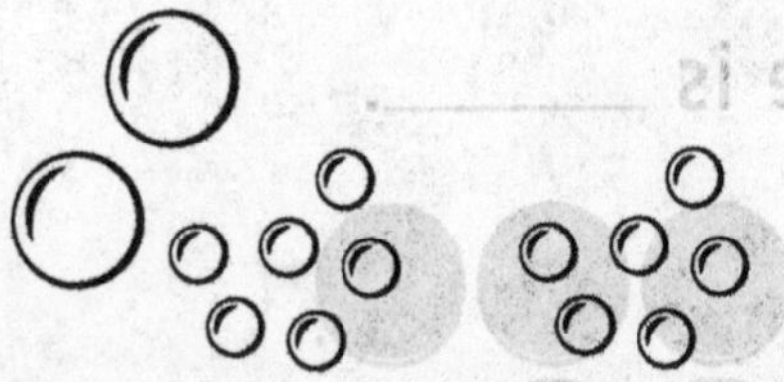

_____ + _____ + _____ = _____

4. Draw 2 sets of bubbles that you blow. How can you add to find out if you win the contest?

Make a 10 to Add

Name ____________________

Review

You can make a 10 to add.

What is 8 + 5?

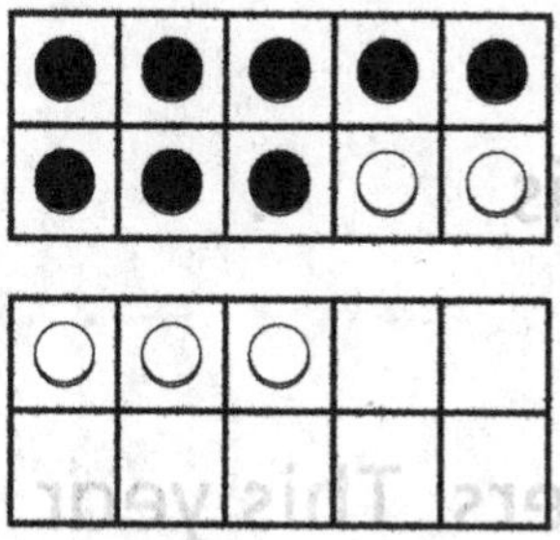

The number **2** makes a 10 with 8.

You can break apart 5 into **2** and **3**.

8 + **2** = 10, and 10 + **3** = 13.

So, 8 + 5 = 13.

How can you make a 10 to add? Write the numbers to show your thinking.

1. 9 + 7

   The number ____ makes a 10 with 9.

   You can break apart 7 into ____ and ____.

   9 + ____ = 10, and 10 + ____ = ____.

2. 6 + 8

   The number ____ makes a 10 with 6.

   You can break apart 8 into ____ and ____.

   6 + ____ = 10, and 10 + ____ = ____.

Make a 10 to Add

Name ______________________

This year new players joined the band. How can you make a 10 to add? Show your thinking.

1. Last year there were 9 flute players. This year 6 more joined. What is the total number of flute players?

   The total number of flute players is _____.

2. Last year there were 7 drum players. This year 4 more joined. What is the total number of drum players?

   The total number of drum players is _____.

3. Last year there were 8 trumpet players. This year 5 more joined. What is the total number of trumpet players?

   The total number of trumpet players is _____.

4. How many flute and trumpet players joined the band this year?

   The total number of flute and trumpet players who joined the band this year is _____.

Choose Strategies to Add

Name ____________________

Review

You can add two numbers in different ways.

You might count on, use doubles, use near doubles, or make a 10. How can you find 6 + 7?

Use near doubles.

6 + 6 = 12

1 more is 13.

Make a 10.

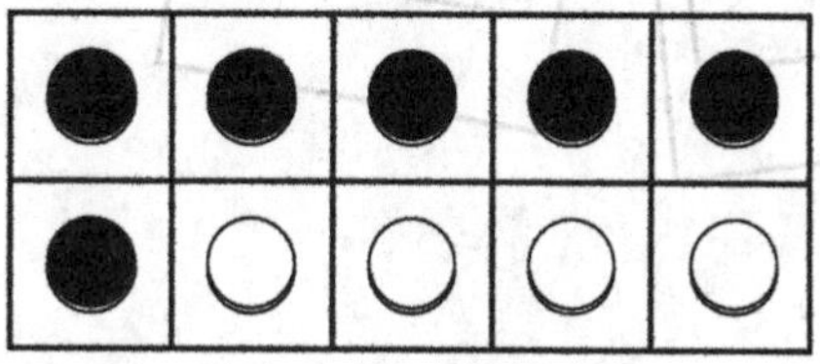

6+4=10

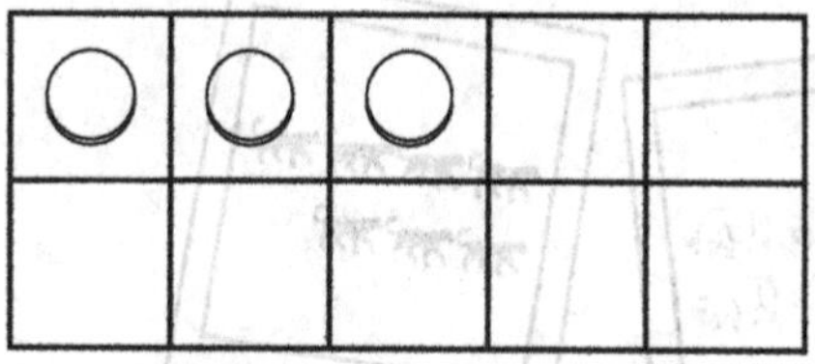

10+3=13

Both strategies help you find the same sum, 13.

What is the sum? Tell or show how you added.

1. 5 + 7 = ____

2. 8 + 7 = ____

3. 6 + 5 = ____

4. 9 + 4 = ____

Choose Strategies to Add

Name ______________________________

Amare took pictures of animal groups. How many animals are in his picture pairs? Explain the strategy you used to find the sum.

1.

2.

3. One of the pictures Amare took is missing. Draw a group of animals that might be missing. How many animals are there? Explain your strategy.

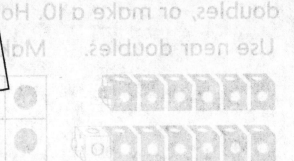

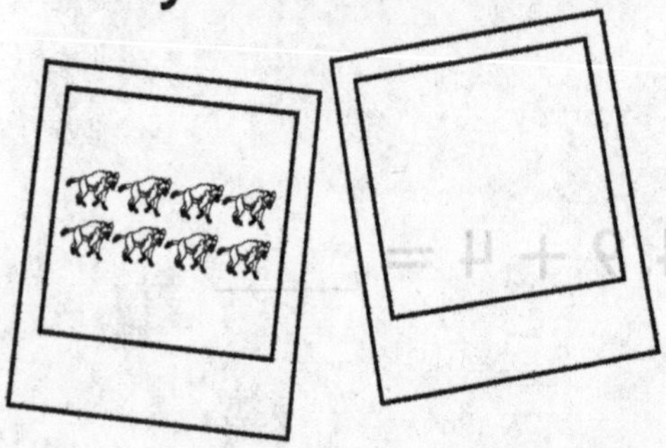

Lesson 4-7 • Reinforce Understanding

Use Properties to Add

Name ______________________________

Review

The order of addends does not change the sum.

9 + 5 = 14

5 + 9 = 14

What is the sum?

1. 6 + 2 = _____　　　2. 4 + 9 = _____
   2 + 6 = _____　　　　9 + 4 = _____

How can you write a different equation with the same addends and sum?

3. 9 + 7 = 16

   _____ + _____ = _____

4. 6 + 5 = 11

   _____ + _____ = _____

Use Properties to Add

Name ______________________________

Use the number of items in the truck as addends. How can you write two different equations using the addends?

1.

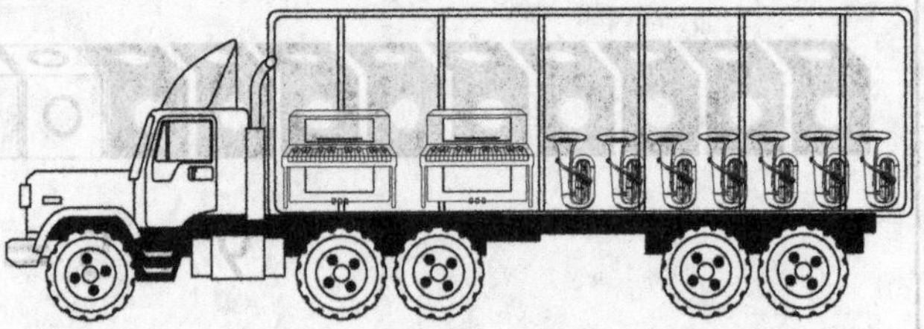

2.

3. Write two different equations with the same addends and a sum of 10. Then write as many more different equation pairs with a sum of 10 as you can.

Add Three Numbers

Name ______________________________

Review

You can add addends in any order. The sum is the same.

What is the sum of 4 + 1 + 3?

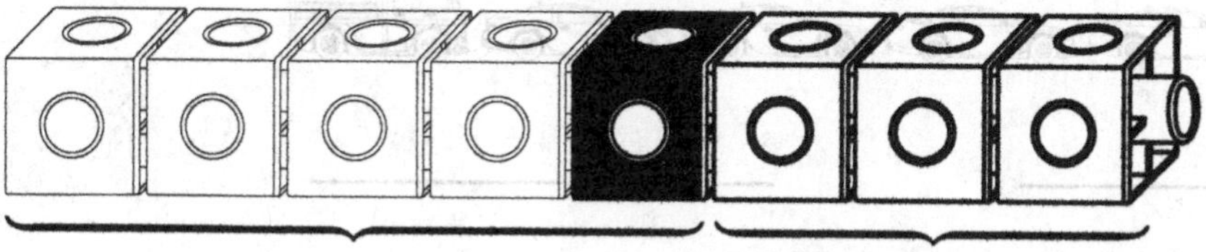

4 + 1 = 5   5 + 3 = 8

3 + 4 = 7   7 + 1 = 8

Add 2 addends. Then add the other addend. What is the sum?

1. 4 + 3 + 7 = ______
2. 8 + 2 + 5 = ______
3. 5 + 6 + 5 = ______
4. 7 + 1 + 9 = ______

What is the sum?

5. 3 + 7 + 7 ______
6. 9 + 8 + 1 ______

Add Three Numbers

Name ______________________________

How can you write 2 different equations to show the number of cars, buses, and trucks?

1. 

______________ ______________

2. 

______________ ______________

3. On another street, there are 12 cars, 2 buses, and 3 trucks. How can you write equations to show the number of cars, buses and trucks? Write as many different equations as you can.

Find an Unknown Number in an Addition Equation

Name ____________________

Review

You can use one addend and the sum to find an unknown addend in an equation.

4 + ? = 10

4 5 6 7 8 9 10

Start with the known addend. Count on by ones until you reach the sum.

The number of counts is the unknown addend.

4 + 6 = 10

What is the unknown addend?

1. 4 + ____ = 11

2. ____ + 2 = 10

3. ____ + 7 = 12

4. 9 + ____ = 15

5. Laura gives some dollars to Carlos. Carlos gets 6 dollars from Ben. Now Carlos has 10 dollars. How many dollars did Laura give Carlos?

Laura gave Carlos ____ dollars.

Find an Unknown Number in an Addition Equation

Name ________________________________

Ravi picks vegetables. Write an addition equation with an unknown addend. Solve.

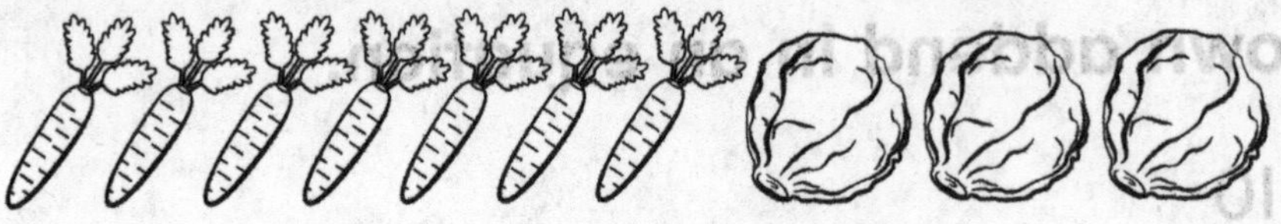

1. Ravi picks 7 carrots. Then he picks some more. Now he has 18 carrots. How many more carrots does he pick?

_____ + _____ = _____

Ravi picks _____ more carrots.

2. Ravi picks 3 cabbages. Then he picks some more. Now he has 12 cabbages. How many more cabbages does he pick?

_____ + _____ = _____

Ravi picks _____ more cabbages.

3. Ravi wants to pick 16 more vegetables for soup. Write a problem about the vegetable he picks. Use an unknown addend. Then write the answer.

Understand the Equal Sign

Name ______________________________

Review

The equal sign (=) means that the amount on the left is the same as the amount on the right.

Sam has 4 blue cars and 7 red cars.

Joey has 2 blue cars and 9 red cars.

Sam　　　　Joey

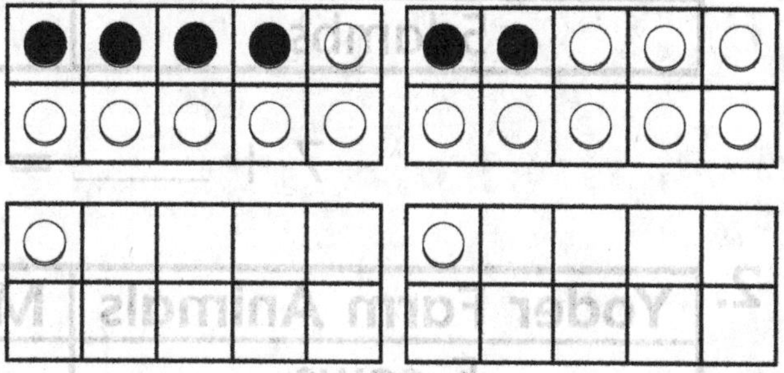

Think:　　$4 + 7 = ?$　　$2 + 9 = ?$

$4 + 7 = 11$　　$2 + 9 = 11$

Both students have an equal number of cars. You can show this with an equal sign.

$4 + 7 = 2 + 9$

How can you fill in the equations to show the sums and equal amounts?

1. $9 + 7 =$ _____　　$8 + 8 =$ _____　　$9 + 7$ _____ $8 + 8$

2. $7 + 7 =$ _____　　$9 + 5 =$ _____　　$7 + 7$ _____ $9 + 5$

3. $8 + 4 =$ _____　　$5 + 7 =$ _____　　$8 + 4$ _____ $5 +$ _____

Understand the Equal Sign

Name ____________________

The Yoder and Miller farms have animals. Fill in the equations to show their numbers of animals.

1.

Yoder Farm Animals	Miller Farm Animals
7 sheep	6 sheep
5 lambs	6 lambs

7 + ____ = ____ + 6

2.

Yoder Farm Animals	Miller Farm Animals
5 cows	8 cows
4 horses	1 horse

5 + ____ = ____ + 1

3.

Yoder Farm Animals	Miller Farm Animals
8 pigs	7 pigs
8 goats	9 goats

8 + ____ = ____ + 9

4. Fill in the sentences. Then write an equation with 6 addends to shows the number of animals.

The Yoders have ____ sheep, 4 horses, and ____ lambs.

The Millers have 9 goats, ____ lambs, and ____ horse.

True Addition Equations

Name ____________________

Review

An equation is true when the amounts on both sides are equal.

Is the equation 4 + 3 = 1 + 6 true or false?

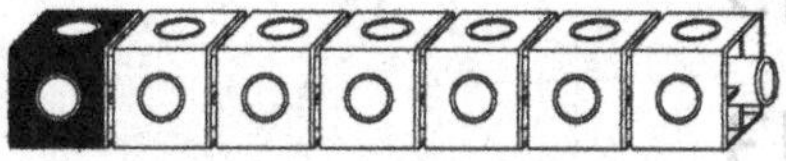

4 + 3 = 7   1 + 6 = 7

Both equal 7, so the equation 4 + 3 = 1 + 6 is true.

Is the equation true? Circle True or False.

1. 8 + 3 = 12

True False

2. 9 + 9 = 17

True False

3. 4 + 8 = 6 + 6

True False

4. 15 = 9 + 6

True False

5. 7 + 6 = 8 + 9

True False

6. 5 + 9 = 6 + 6 + 2

True False

True Addition Equations

Name ____________________

You are checking your work. If the equation is true, circle it. If the equation is false, make it true. Rewrite it correctly with a different sum.

1. 9 + 7 = 16

2. 6 + 8 = 15

3. 5 + 8 = 13

4. 18 = 14 + 6

5. 19 = 7 + 11

6. 7 + 10 = 17

7. 5 + 9 = 6 + 7 is false. Show 2 different ways you could change the addends to make it true.

8. 8 + 4 + 7 = 10 + 1 + 7 is false. Show 3 different ways you could change the addends to make it true.

Relate Counting to Subtraction

Name ____________________

Review

You can count or subtract to find a difference.

10 − 4 = ?

Use 10 connecting cubes. Then cross out 4 cubes.

Count the cubes that are left.

10 − 4 = 6

What is the difference? Use connecting cubes.

1. 8 − 3 = ____

2. 6 − 2 = ____

3. 9 − 4 = ____

4. 10 − 7 = ____

What is the difference?

5. 12 − 5 = ____

6. 5 − 1 = ____

Relate Counting to Subtraction

Name ______________________

The table shows how many game tokens three friends bring to a family fun center.

Name	Tokens
Niles	8
Trey	11
Ellie	7

Use the information in the table. Solve.

1. Niles plays the claw machine game. He spends 3 tokens. How many does he have left?

   _____ tokens

2. Trey spends 4 tokens playing video games. How many does he have left?

   _____ tokens

3. Ellie gives 2 tokens to her sister. How many does Ellie have left?

   _____ tokens

4. The three friends combine the tokens they have left. How many tokens do they have now? Explain your thinking.

Count Back to Subtract

Name ______________________________

Review

You can count back to subtract.

8 − 5 = ?

Start at 8. Count back by 5. You end at 3.

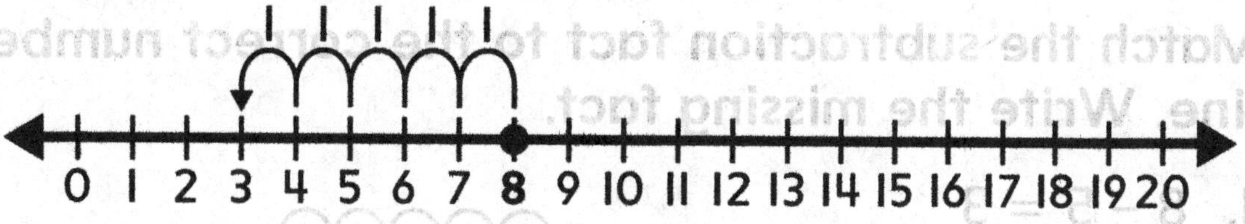

8 − 5 = 3

Start at 8. Count back to 5. There are 3 jumps.

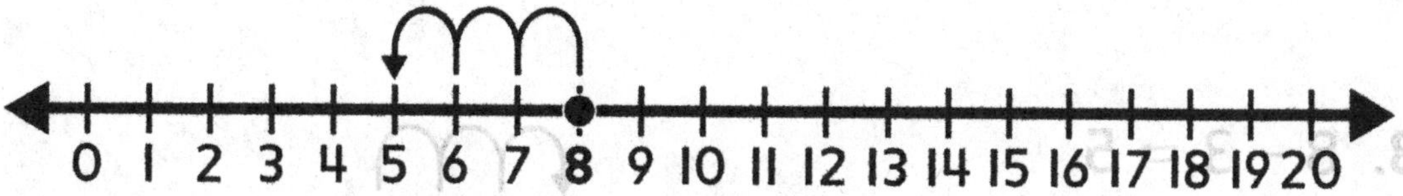

8 − 5 = 3

What is the difference?

1. 10 − 6 = ____

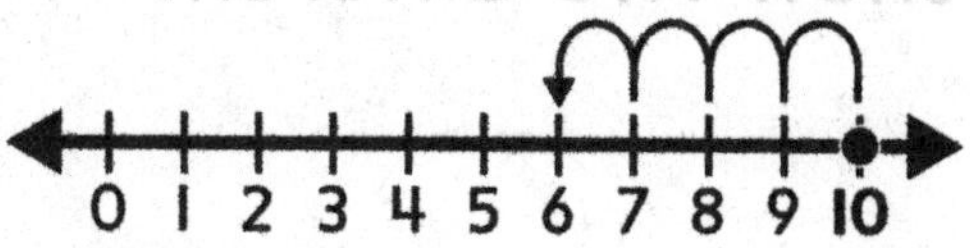

2. 6 − 2 = ____

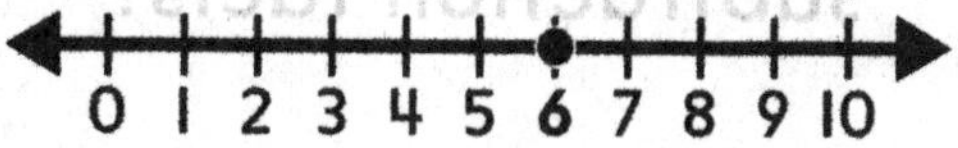

What is the difference?

3. 10 − 8 = ____

4. 7 − 4 = ____

Count Back to Subtract

Name ______________________________

A number line can show two subtraction facts.

This number line shows 10 – 6 = 4 and 10 – 4 = 6.

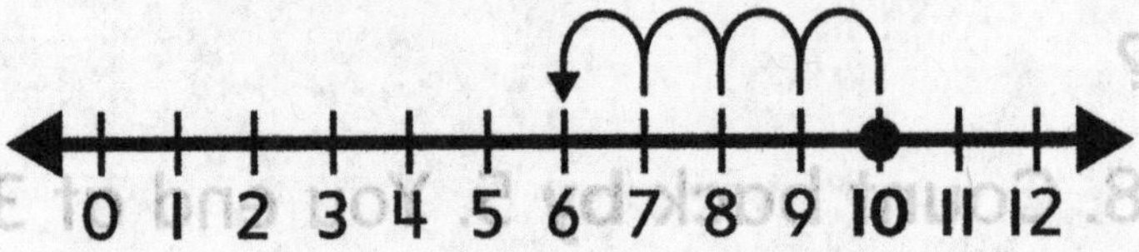

Match the subtraction fact to the correct number line. Write the missing fact.

1. 8 – 5 = 3

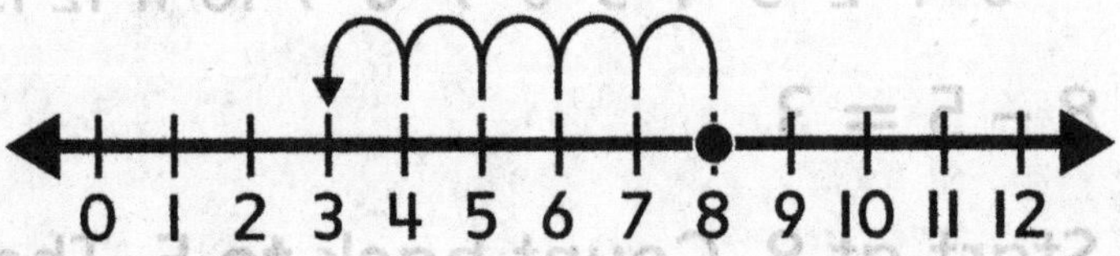

2. 7 – 3 = 4

3. 8 – 3 = 5

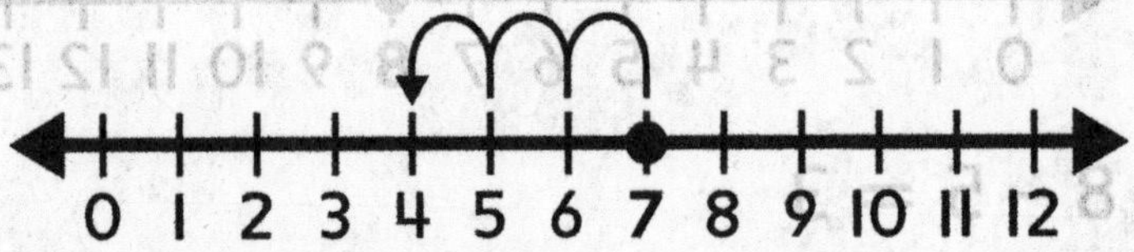

4. ____________

5. How can a number line show two different subtraction facts?

Count On to Subtract

Name ______________________

Review

You can count on to subtract.

Start at the lesser number. Count on to the total.
The difference is the number you counted.

8 – 3 = ?

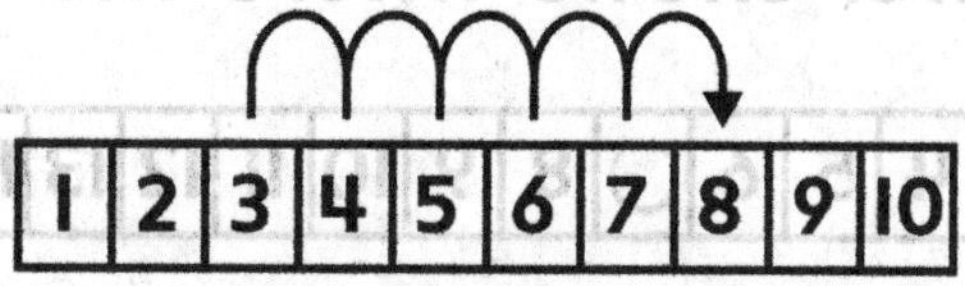

There are 5 counts. The difference is 5.

8 – 3 = 5

What is the difference? Use a number path.

1. 7 – 5 = ____

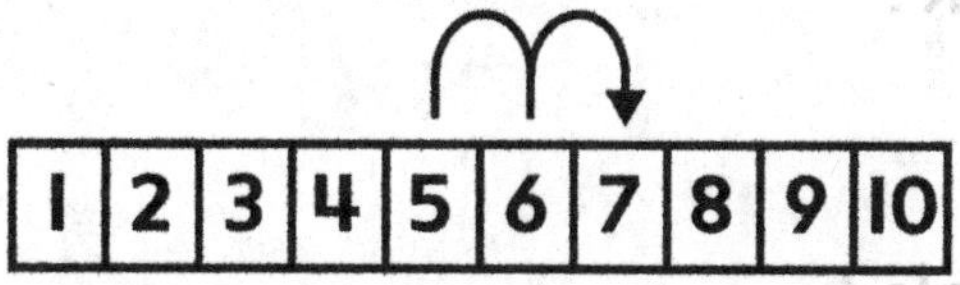

2. 9 – 3 = ____

1	2	3	4	5	6	7	8	9	10

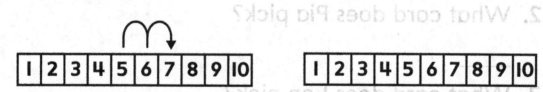

What is the difference?

3. 8 – 4 = ____

4. 9 – 7 = ____

5. 6 – 4 = ____

6. 10 – 3 = ____

Count On to Subtract

Name ______________________________

In the game, *Closest to One,* students pick a number card from 1 to 10. Then they move backward along a number path that many spaces. The first player to land exactly on 1 wins.

The dark counter shows where the student starts. The light counter shows where the student ends.

Anita	1	2	3	4	5	6	○	8	9	10	11	12	13	14	15	16	●	18	19	20
Pia	1	2	3	4	5	○	7	8	9	10	11	12	13	14	●	16	17	18	19	20
Len	1	2	3	4	5	6	7	8	9	10	11	○	13	14	15	16	17	●	19	20

1. What card does Anita pick?

   ______

2. What card does Pia pick?

   ______

3. What card does Len pick?

   ______

4. Which of the players can win with their next pick? Why?

Make a 10 to Subtract

Name ____________________

Review

You can make a 10 to subtract.

13 − 6 = ?

Make a row of 13 connecting cubes.

Take away 3 cubes to make a 10

Take away 3 more cubes so a total of 6 is taken away.

The difference is 7.

10 − 3 = 7

What is the difference? Use connecting cubes.

1. 12 − 7 = ____

2. 14 − 8 = ____

What is the difference?

3. 15 − 6 = ____ **4.** 11 − 3 = ____

Make a 10 to Subtract

Name ______________________________

The table shows how many stickers Mrs. Alvarez has at the start of the day to give to her students.

Sticker	Total
Star	18
Rainbow	15
Soccer	17

Use the information in the table. How can you make a 10 to subtract? Solve.

1. Mrs. Alvarez gives 9 star stickers to students. How many are left?

   ____ star stickers

2. Mrs. Alvarez gives 6 rainbow stickers to Marci. She gives 2 rainbow stickers to Karla. How many rainbow stickers are left?

   ____ rainbow stickers

3. Mrs. Alvarez gives 9 soccer stickers to students. How many are left? Explain your thinking.

   ____ soccer stickers

Use Near Doubles to Subtract

Name ______________________________

Review

You can use doubles and near doubles facts to subtract.

16 − 7 = ?

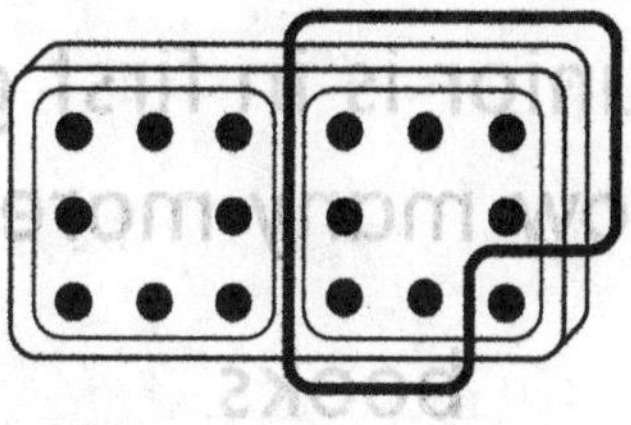

8 + 8 = 16, so 16 − 8 = 8

You take away 1 less in 16 − 7 than in 16 − 8.

So the difference of 16 − 8 is 1 more than 8.

16 − 7 = 9

What is the difference? Write the numbers.

1. 14 − 7

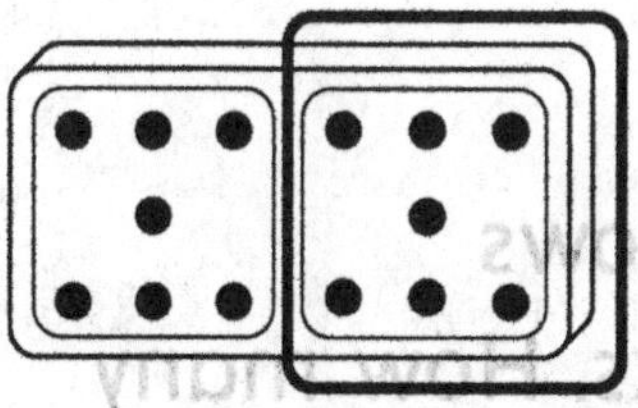

7 + 7 = ____

14 − 7 = ____

2. 10 − 6

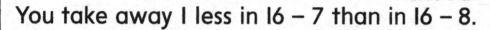

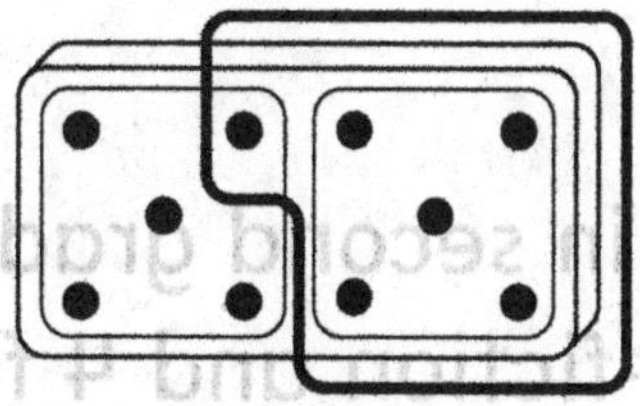

5 + 5 = ____

10 − 5 = ____

10 − 6 is ____ less than 10 − 5

10 − 6 = ____

3. 16 − 8 = ____

4. 12 − 5 = ____

Use Near Doubles to Subtract

Name ______________________________

First graders can borrow 10 library books at Spring School. Second graders can borrow 17. Use doubles or near doubles facts to help you.

1. Junior is in first grade. He borrows 5 books. How many more books can he borrow?

   _____ books

2. Karen is in Junior's class. She borrows 1 less book than Junior. How many more books can she borrow?

   _____ books

3. Yin is in second grade. She borrows 5 non-fiction and 4 fiction books. How many more books can she borrow?

   _____ books

Use Addition to Subtract

Name ____________________

Review

You can use addition to help you subtract.

9 − 4 = ?

Make an addition equation with an unknown addend.

4 + ? = 9

Use an addition fact you know to find the unknown addend and difference.

4 + **5** = 9, so 9 − 4 = **5**

What is the difference? Use the addition equation to subtract.

1. 5 + 3 = 8, so

8 − 5 = ____

2. 6 + 7 = 13, so

13 − 6 = ____

3. 6 + ____ = 11, so

11 − 6 = ____

4. 4 + ____ = 16, so

16 − 4 = ____

5. 8 + ____ = 12, so

12 − 8 = ____

6. 1 + ____ = 10, so

10 − 1 = ____

Use Addition to Subtract

Name ____________________

Complete the addition equation. Then match the subtraction problem to an addition equation to help you subtract. Then write the difference.

1. 13 − 7 = ____ 8 + ____ = 13

2. 10 − 6 = ____ 3 + ____ = 10

3. 13 − 8 = ____ 5 + ____ = 15

4. 15 − 5 = ____ 7 + ____ = 13

5. 10 − 3 = ____ 6 + ____ = 10

6. 18 − 3 = ____ 3 + ____ = 18

7. Samir knows that 9 + 8 = 17. How can he use this fact to help him subtract 17 − 9? Explain.

Use Fact Families to Subtract

Name ______________________________

Review

You can use a fact family to complete an equation.

15 − ? = 6

A number bond can show a fact family.

9 + 6 = 15

6 + 9 = 15

15 − 6 = 9

15 − 9 = 6

So, 15 − 9 = 6.

Write the fact family for the number bond.

1.

4 + ____ = 11

7 + 4 = ____

11 − ____ = 7

____ − 7 = 4

2.

____

____

____

____

Use Fact Families to Subtract

Name ______________________________

Could the fact be part of the fact family on the fact triangle? Write Yes or No. If the answer is no, explain your thinking.

1. 9 + 5 = 14

   ______

2. 9 − 5 = 4

   ______

3. 4 + 14 = 18

   ______

4. 14 − 5 = 9

   ______

5. The greatest number in a fact triangle is 13. How can you make three different fact triangles? Make the triangles.

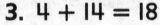

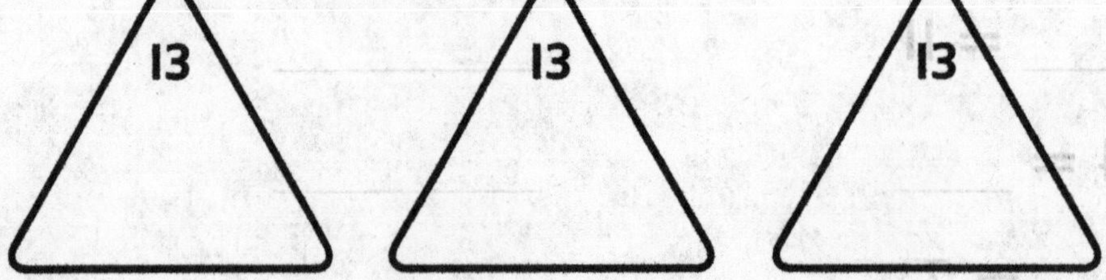

Find an Unknown in a Subtraction Equation

Name ____________________

Review

You can find an unknown in different ways.

13 − ? = 9

One way is to use a fact family.

4 + 9 = 13   9 + 4 = 13

13 − 9 = 4   13 − 4 = 9

So, 13 − 4 = 9.

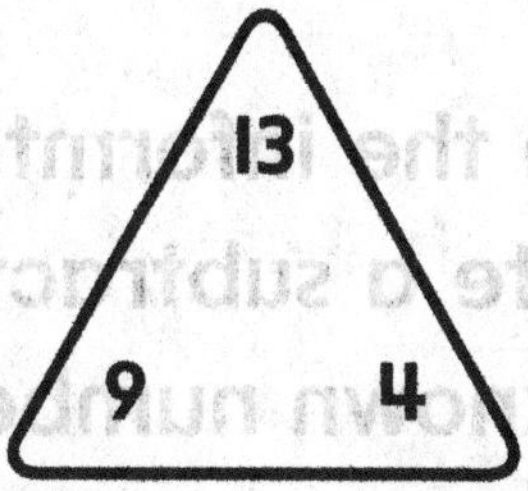

How can you complete the equation?
Tell or show how you solved.

1. 10 − ____ = 4

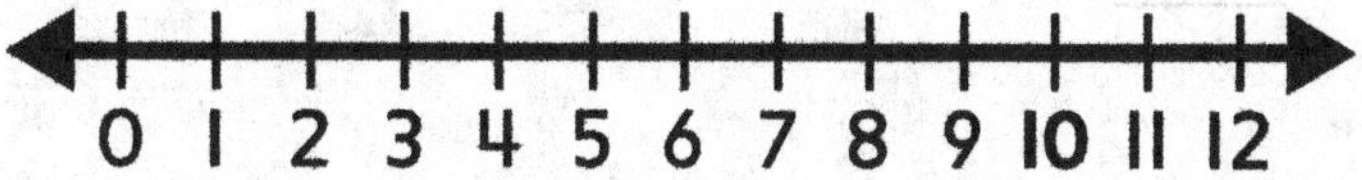

2. ____ − 10 = 7

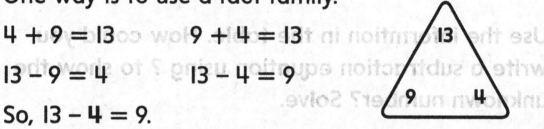

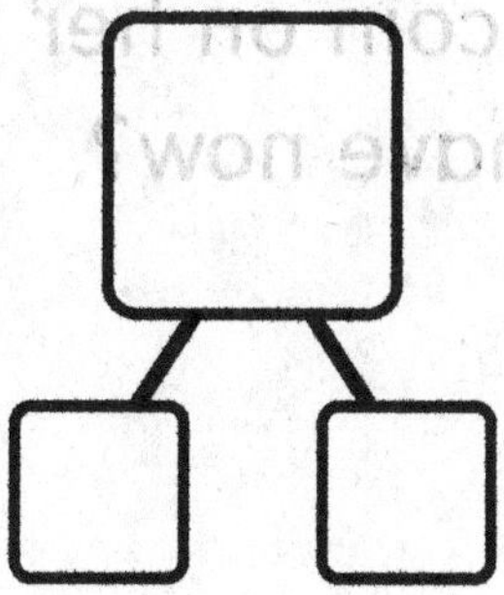

3. 14 − ____ = 5   **4.** 7 = 12 − ____

Lesson 5-8 • Extend Thinking

Find an Unknown in a Subtraction Equation

Name ____________________

Some friends are playing a board game. The table shows how many coins each player has at the start of their turns.

Name	Coins
Tyrese	8
Antonia	12
Theo	7

Use the informtion in the table. How could you write a subtraction equation using ? to show the unknown number? Solve.

1. Tyrese has 11 coins after his turn. How many coins does he earn on his turn?

____ coins

2. Antonia earns 1 coin on her turn. How many coins does she have now?

____ coins

3. Theo has 12 coins after his turn. How many coins does he earn on his turn?

____ coins

True Subtraction Equations

Name ________________________________

Review

An equation is true when the amounts on both sides are equal.

$8 - 3 \stackrel{?}{=} 10 - 5$

Each row has 5 remaining. This equation is true.

Is the equation true? Circle True or False.

1. $7 - 3 \stackrel{?}{=} 9 - 4$

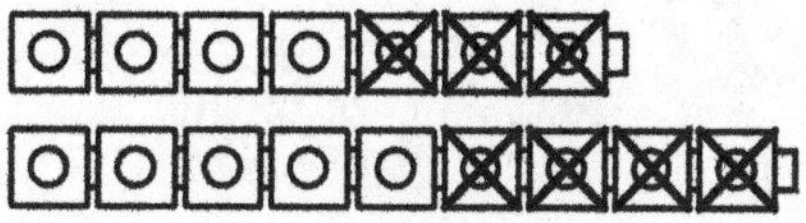

True   False

2. $8 - 4 \stackrel{?}{=} 6 - 2$

True   False

3. $12 - 3 \stackrel{?}{=} 9 - 1$

True   False

4. $9 - 3 \stackrel{?}{=} 10 - 4$

True   False

True Subtraction Equations

Name ____________________

Each equation is false. How could you change the representation and equation to make it true?

1. $9 - 3 \overset{?}{=} 11 - 4$

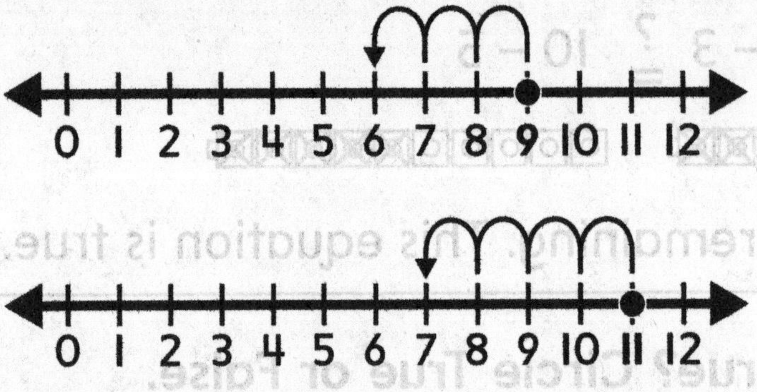

2. $8 - 3 \overset{?}{=} 10 - 4$

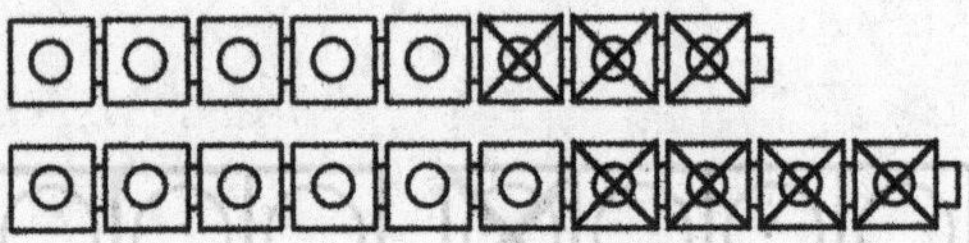

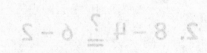

3. $6 - 2 \overset{?}{=} 9 - 6$

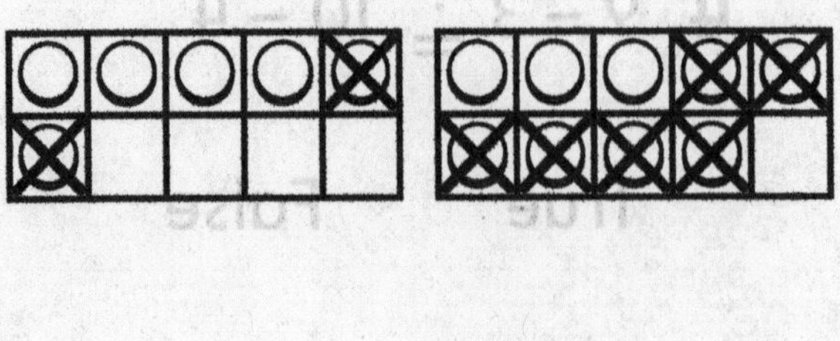

Understand Defining Attributes of Shapes

Name ______________________________

Review

You can describe rectangles and squares by their number of vertices and sides.

rectangle

4 vertices

4 sides

Short sides same length.

Long sides same length.

square

4 vertices

4 sides

All sides same length.

The shapes are closed. Their vertices are the same.

1. Complete the chart.

	Vertices	Sides	What Else?
rectangle			
square			

Draw an example of the closed shape.

2. square

3. rectangle

Understand Defining Attributes of Shapes

Name ____________________

How do you know the shape is *not* a rectangle? Match the shape with one reason only.

1. 

2. 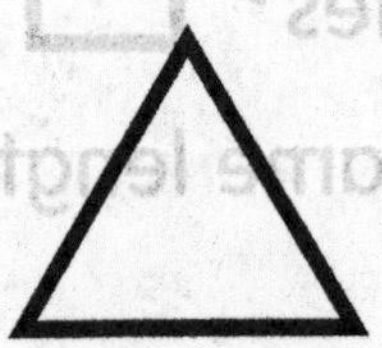

3. 

4. 

5. 

Shape is not closed.

Shape has more than 4 sides.

Shape is not 2-dimensional.

Shape has different vertices.

Shape has fewer than 4 sides.

Understand Non-Defining Attributes

Name ______________________________

Review

You can describe 2-dimensional closed shapes.

Words that Help Name the Shape	Words that Do *Not* Help Name the Shape
number of vertices number of sides lengths of sides	color size position

Match the description with the closed shape.

1. 6 sides

not enough information

2. big and purple

square

3. 4 sides the same length and 4 vertices the same

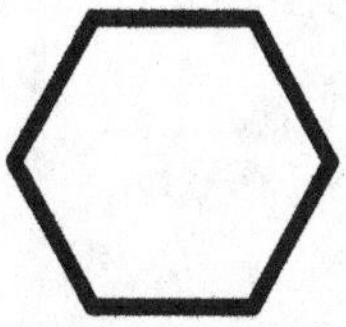

hexagon

Understand Non-Defining Attributes

Name ________________________________

Use the instructions to draw a house.

1. Draw a large square for the front of the house. Color it red.
2. Draw a long triangle for the roof. Color it gray.
3. Draw a tall rectangle for the door. Color it brown.
4. Draw a small circle for the doorknob. Color it green.
5. Draw 2 hexagons for the windows. Color them yellow.
6. Does coloring the shapes change what the shapes are? Explain.

Compose Shapes

Name ______________________

Review

Shapes can be put together to make either another shape or a design.

This hexagon is made from 6 triangles.

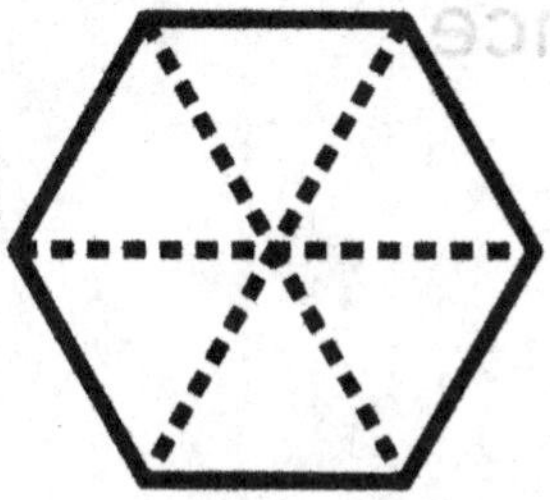

This cat is made from many different shapes.

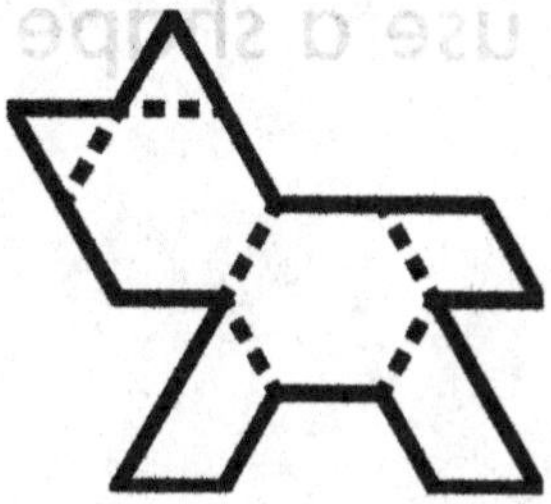

Can the  **be made from the shapes shown? Circle Yes or No.**

1.  

Yes No

2.  

Yes No

3. Draw a large triangle using some of these shapes.

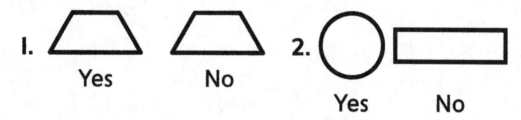

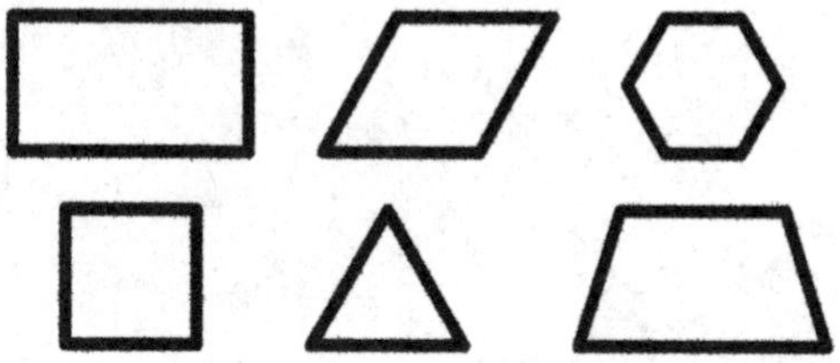

Compose Shapes

Name ______________________________

Use the shapes shown.

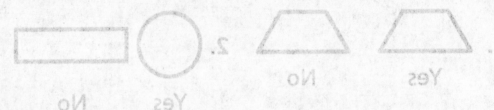

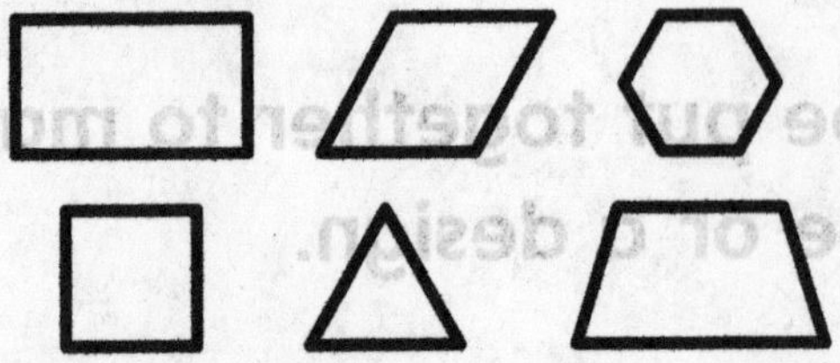

1. Draw how to make an animal using the shapes. You may use a shape more than once.

2. Draw how to make a robot using the shapes. You may use a shape more than once.

Build New Shapes

Name ______________________________

Review

A shape can be taken apart to make a different shape.

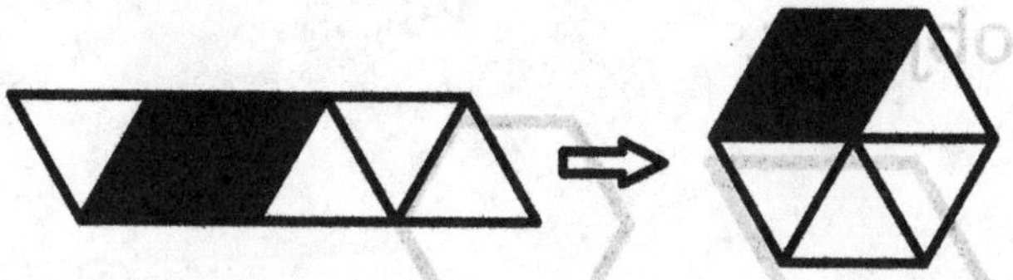

Can the parts of Shape A be used to make Shape B? Explain your answer.

1.

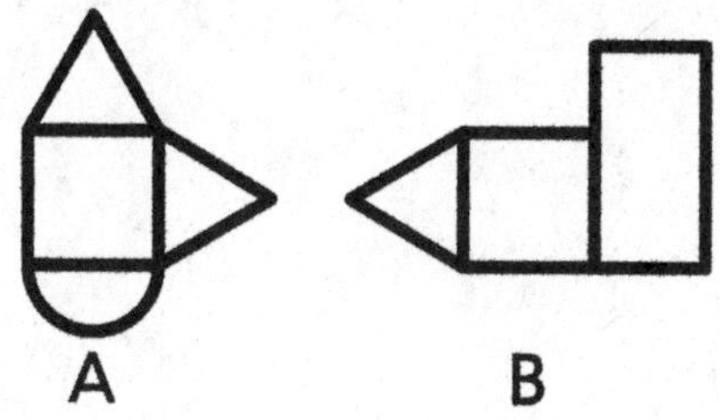

2.

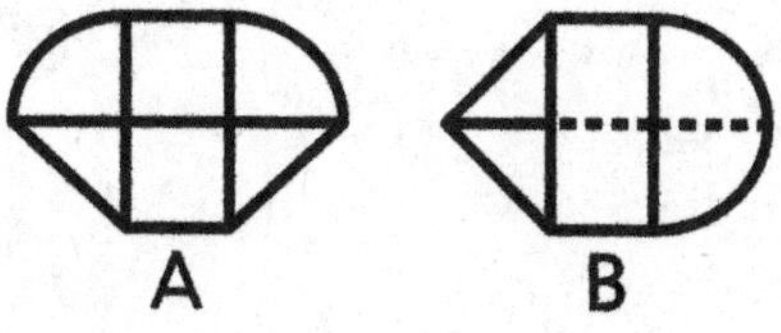

Build New Shapes

Name ______________________________

I. How can you use the shapes below to make an object? Use at least 2 different shapes, and a total of at least 3 shapes. Then take your object apart and use the parts to draw a different object. Name both objects.

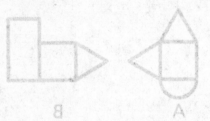

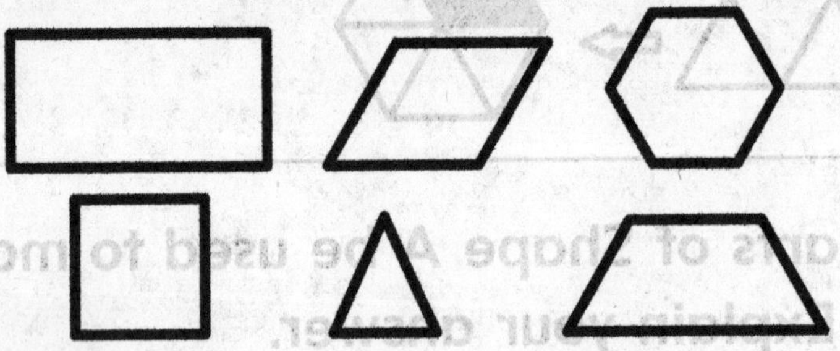

Understand Attributes of Solids

Name ____________________

Review

You can describe 3-dimensional shapes using words.

Words that Help Name the Shape	Words that Do *Not* Help Name the Shape
number of vertices number of faces number of edges	color size position

Match the description with the shape.

1. I vertex and I face

2. 8 vertices, 6 square faces, and 12 edges

3. small and yellow

not enough information

cone

cube

Understand Attributes of Solids

Name ______________________________

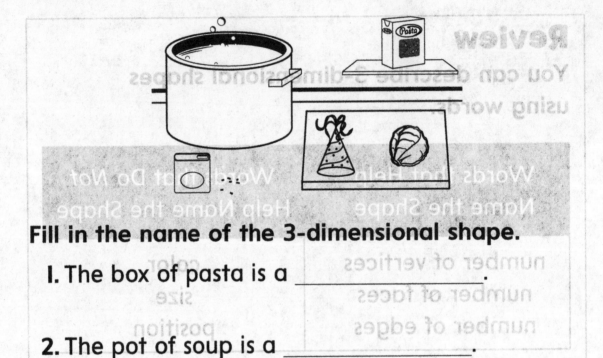

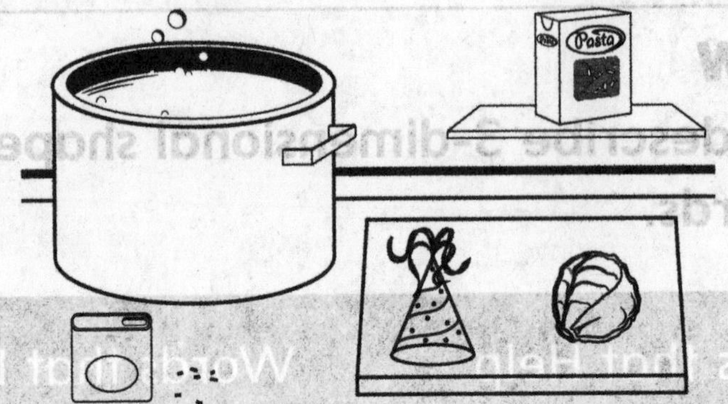

Fill in the name of the 3-dimensional shape.

1. The box of pasta is a ______________.

2. The pot of soup is a ______________.

3. The pepper shaker is a ______________.

4. The head of cabbage is a ______________.

5. The party hat is a ______________.

6. What 3 items around you are 3-dimensional shapes? List the 3 items and name the shapes.

Build New Solids

Name ______________________________

Review

You can put 3-dimensional shapes together to make new shapes.

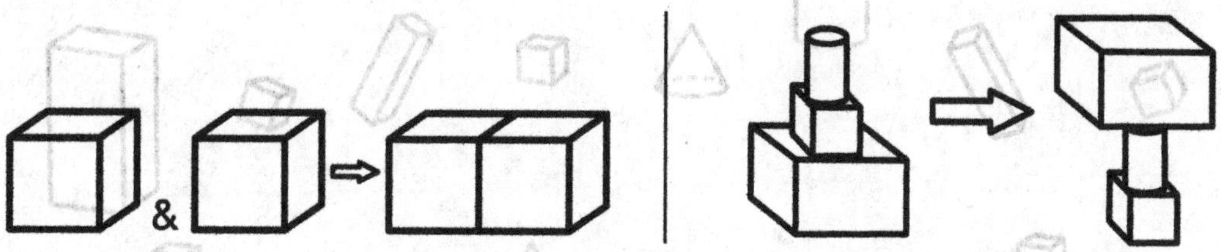

Put all the shapes together. Circle the shape they can make.

1. 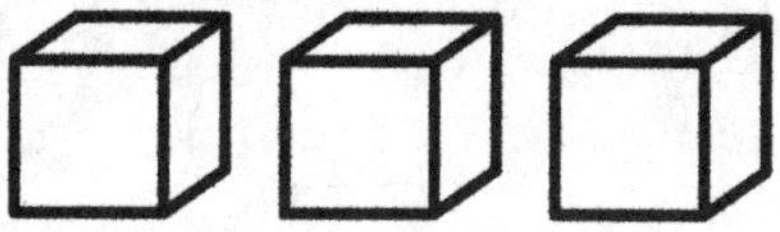 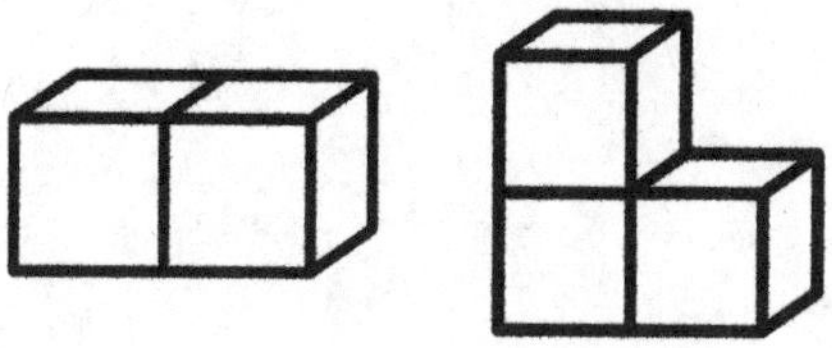

2. 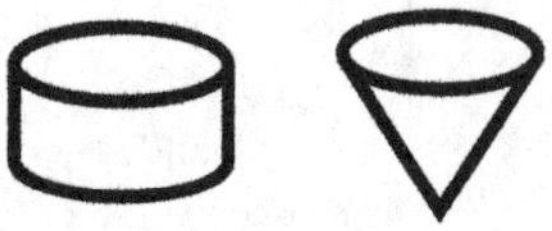 

Can you make Shape B from Shape A? Circle Yes or No.

3. 

Yes   No

4. 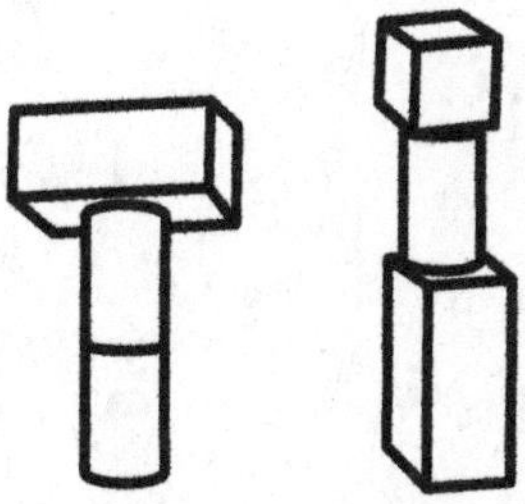

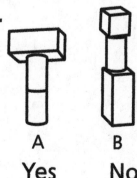

Yes   No

Build New Solids

Name ______________________________

Asli built a robot out of blocks, but her cat knocked it over. Draw a robot using her blocks. Name each shape in your drawing.

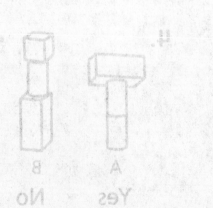

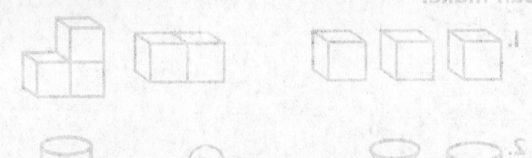

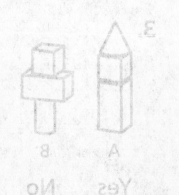

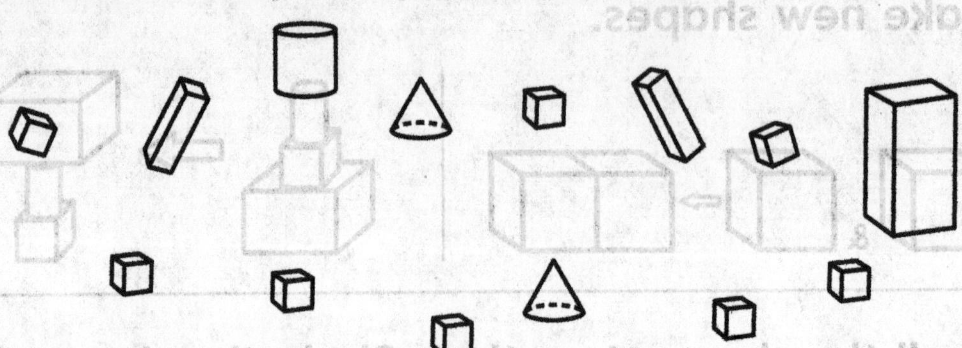

Represent and Solve Add To Problems

Name ______________________________

Review

You can add one addend to another addend to find the unknown sum.

There are 4 children at a table. 3 more children join. How many children are at the table now?

You can use a number line to solve the problem.

4 + 3 = ?

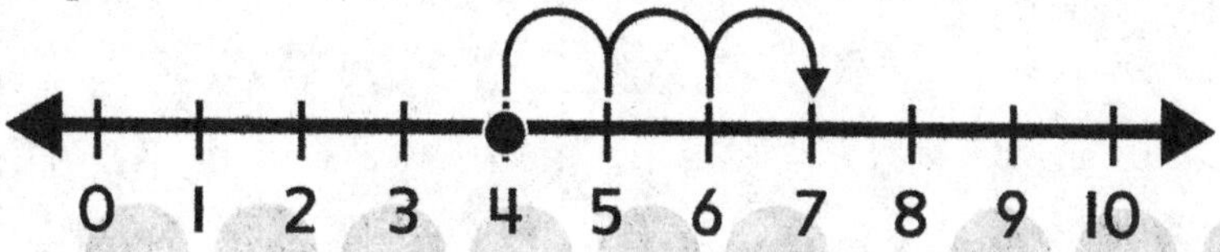

4 + 3 = 7

There are 7 children at the table.

1. How can you make an equation to show the problem? Use ? for the unknown. Use the number line to solve.

   There are 5 birds on a feeder. 9 more birds join them. How many birds are on the feeder now?

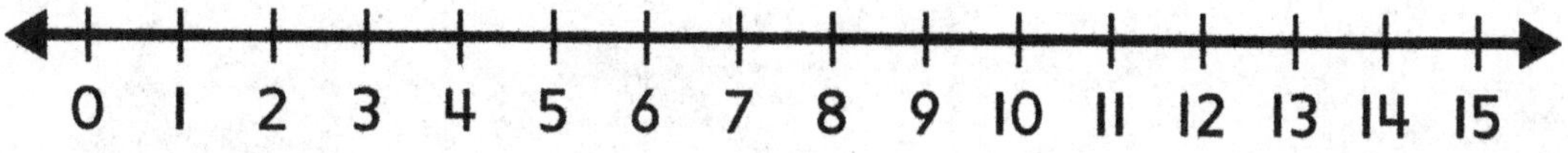

______ birds

Represent and Solve Add To Problems

Name ______________________________

How can you write an addition word problem with an unknown sum to match the picture? Solve the problem.

1. 

2. 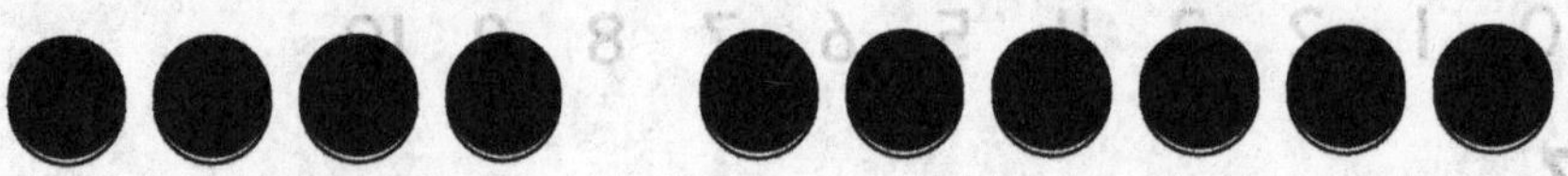

3. 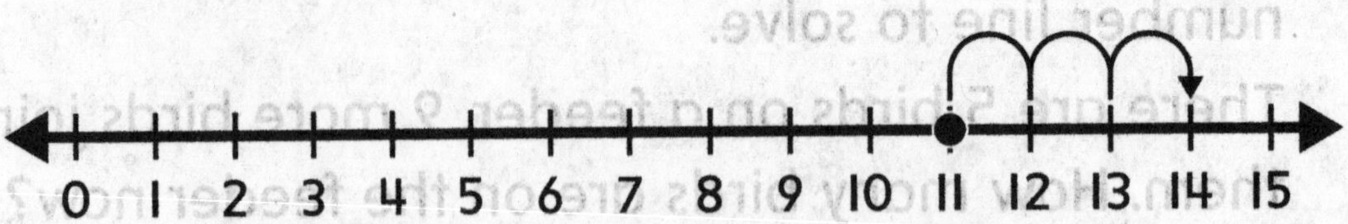

Represent and Solve More Add To Problems

Name ______________________

Review

You can use the known addend and sum to find the unknown addend.

Kim has 8 stickers. How many more stickers does she need for a total of 12 stickers?

8 + ? = 12

Start with 8 connecting cubes. Add cubes until you get 12 total.

8 + 4 = 12

Kim needs 4 more stickers.

1. How can you make an equation to show the problem? Use ? for the unknown. Draw connecting cubes to show your thinking. Solve.

Joe has some coins. Sal gives him 7 more coins. Now Joe has 15 coins. How many coins did Joe have to start?

______ coins

Represent and Solve More Add To Problems

Name ______________________________

The table shows the points 3 friends on a basketball team score in each half of a game.

Player	Half 1	Half 2	Total
Kyle	5	?	9
Jenna	?	6	12
Vince	1	?	10

Use the information in the table. How can you draw or make an equation to show the unknown addend problem? Solve.

1. How many points does Kyle score in Half 2?

_____ points

2. How many points does Jenna score in Half 1?

_____ points

3. How many points does Vince score in Half 2?

_____ points

Represent and Solve Put Together Problems

Name ______________________________

Review

You can add two addends, or parts, together to find the unknown sum, or whole.

There are 8 boys and 6 girls on the playground. How many children are on the playground?

$$8 + 6 = ?$$

You can use connecting cubes to help you solve the problem. Start with 8 cubes. Add 6 cubes.

$8 + 6 = 14$

There are 14 children on the playground.

1. How can you make an equation to show the problem? Use ? for the unknown. Draw connecting cubes to show your thinking. Solve.

   Antonio has 5 pears. Zac has 7 pears. How many pears do they have in all?

_______ pears

Represent and Solve Put Together Problems

Name ______________________________

How can you write an addition word problem with an unknown sum to match the picture? Solve the problem.

1.

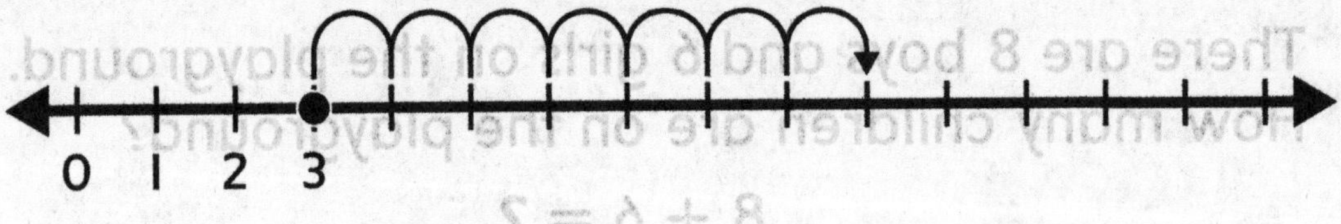

2.

3.

Part	Part
3	15
Whole	

Represent and Solve More Put Together Problems

Name ______________________________

Review

You can use numbers in a word problem to find unknown addends.

Tien has 16 socks in her drawer. She has 10 white socks. The rest are black. How many black socks does she have?

10 + ? = 16

You can use a number bond to help you solve the problem.

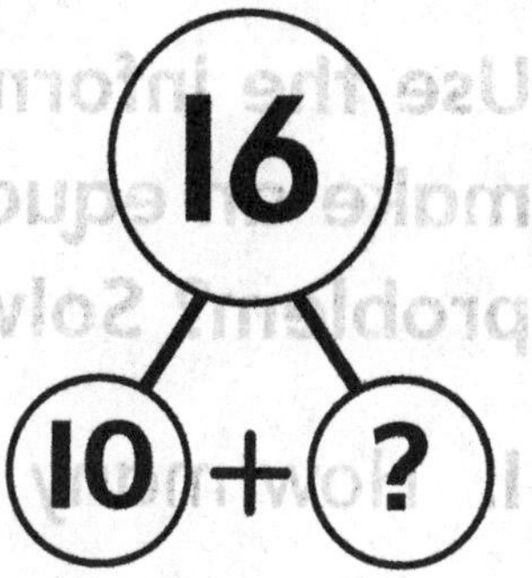

10 + **6** = 16

Tien has 6 black socks.

1. How can you make an equation to show the problem? Use ? for the unknown. Use the number bonds to help you. Solve.

A store has 19 apples. There are 7 red apples. The rest are green. How many green apples are there?

______ green apples

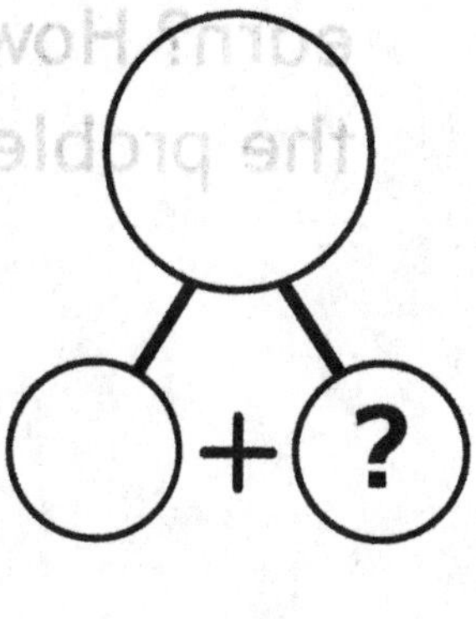

Represent and Solve More Put Together Problems

Name ______________________________

The table shows the number of stickers 3 students earn.

Student	Star	Smile	Total
William	?	9	15
Regina	11	?	17
Anika	?	?	14

Use the information in the table. How can you make an equation to show the unknown addend problem? Solve.

1. How many star stickers does William earn?

    ______ star stickers

2. How many smile stickers does Regina earn?

    ______ smile stickers

3. How many star and smile stickers does Anika earn? How many different ways could you solve the problem? Solve in 2 ways.

Represent and Solve Addition Problems with Three Addends

Name ______________________________

Review

You can add three addends to find the unknown sum.

There are 5 turtles on a log. There are 3 turtles on another log. There are 6 turtles on another log. How many turtles are there in all?

$5 + 3 + 6 = ?$

You can use counters to solve the problem.

●●●●● ●●● ●●●●●●

$5 + 3 + 6 = 14$

There are 14 turtles in all.

1. How can you make an equation to show the problem? Use ? for the unknown. Draw counters to show your thinking. Solve.

   There are 3 red crayons, 6 green crayons, and 4 blue crayons in a box. How many crayons are there?

   _____ crayons

Represent and Solve Addition Problems with Three Addends

Name ______________________________

The picture shows the starting point and locations of clues for a scavenger hunt. Each distance is measured in steps. Use the information in the picture.

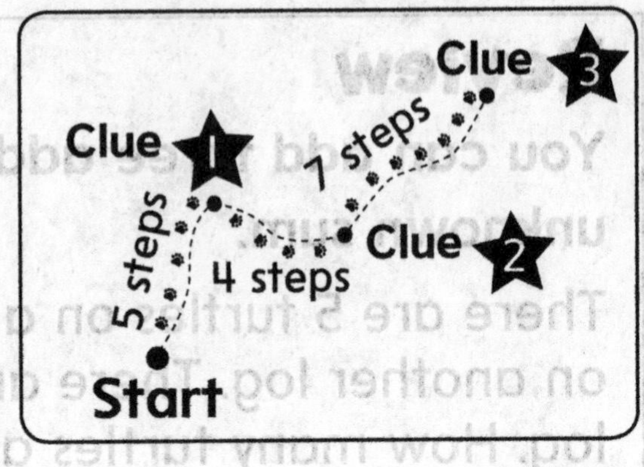

1. Trey wants to know the total distance from the Start to Clue 3 along the path. How can you make an equation using ? for the unknown to show the problem?

2. How can you make a drawing to help you solve the addition problem?

3. What is the total distance from the Start to Clue 3 along the path?

   _____ steps

Solve Addition Problems

Name ____________________

Review

You can find unknown addends or an unknown sum in word problems.

There are 11 birds. 6 are cardinals and the rest are robins. How many robins are there?

$6 + ? = 11$

You can use a number line to help you solve the problem. Start at 6. Count on 5 jumps.

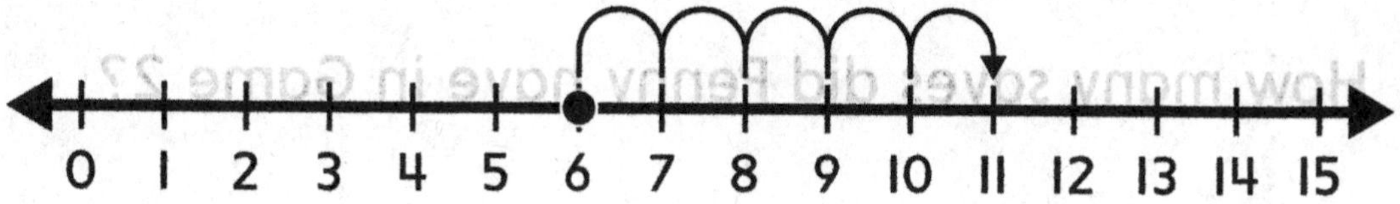

$6 + 5 = 11$

There are 5 robins.

1. How can you make an equation to show the problem? Use ? for the unknown. Use the number line to solve.

   There are 5 adult bunnies and 9 baby bunnies in a field. How many total bunnies are there?

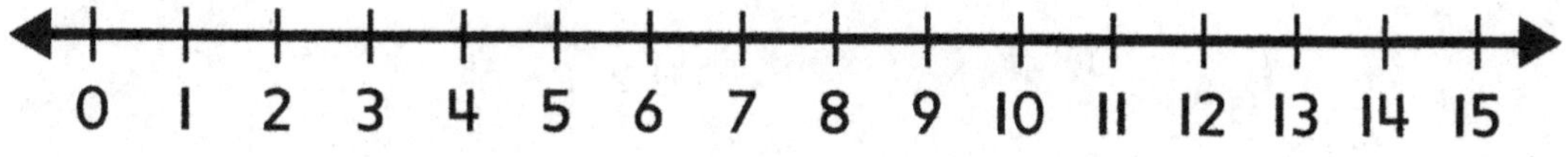

_____ bunnies

Solve Addition Problems

Name ____________________

The table shows the saves made by soccer goalies for 2 games.

Player	Game 1	Game 2	Saves
Penny	11	?	19
Mak	?	6	15
Fran	?	?	12

Use the information from the table. How can you make an equation to show the problem? Solve.

1. How many saves did Penny have in Game 2?

   ______ saves

2. How many saves did Mak have in Game 1?

   ______ saves

3. How many saves did Fran have in each game? Solve in two different ways.

Represent and Solve Take From Problems

Name ____________________

Review

You can subtract one part from the total to find an unknown part or difference.

There are 14 ducks. 6 fly away. How many are left?

$14 - 6 = ?$

You can use a number line to solve the problem.

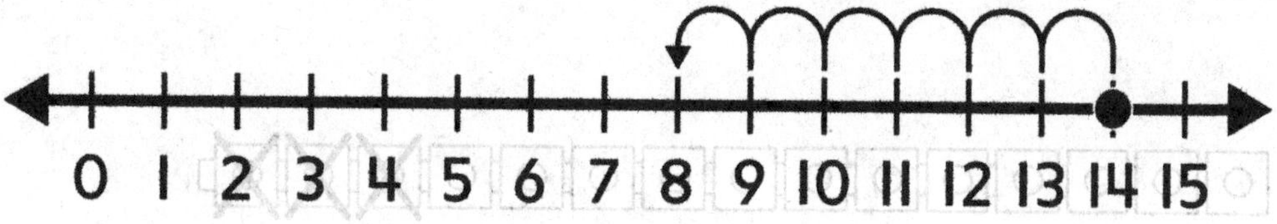

$14 - 6 = \mathbf{8}$

There are 8 ducks left.

1. How can you make an equation? Use ? for the unknown. Use the number line to solve.

   Marty buys 15 game tickets. He uses 9 game tickets. How many tickets does he have left?

   Equation:

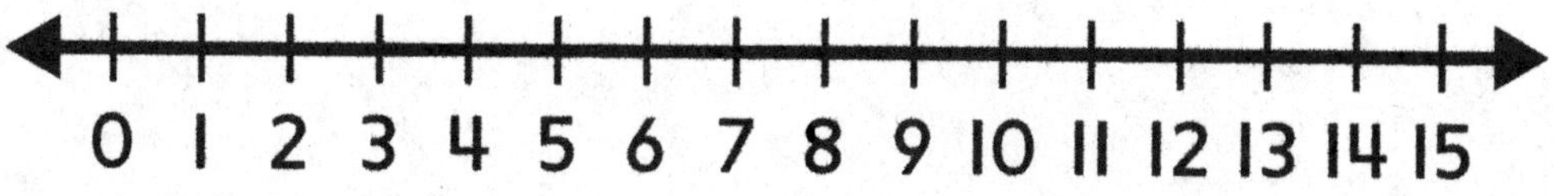

______ tickets

Represent and Solve Take From Problems

Name ______________________________

How can you write a subtraction word problem with an unknown difference to match the picture? Solve the problem.

1. 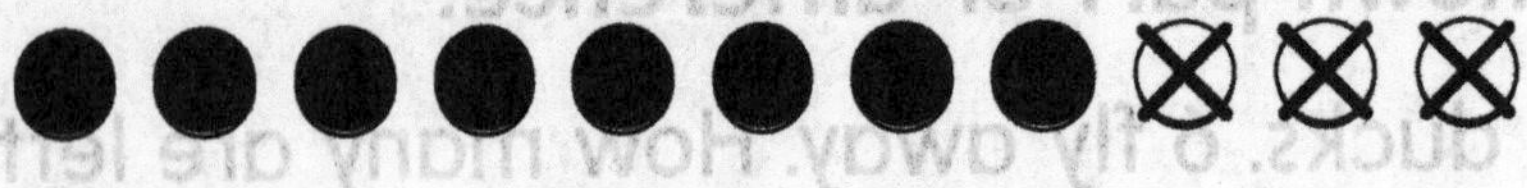

2. 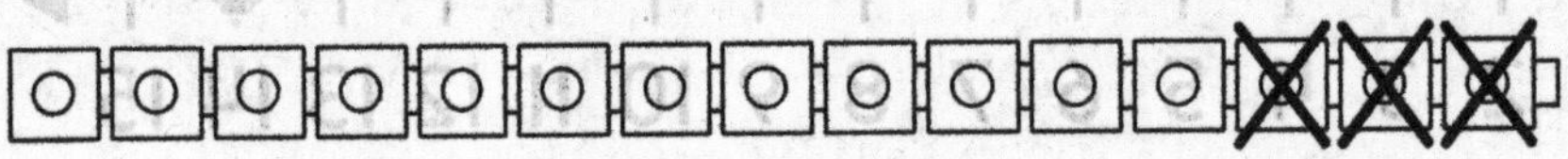

3. 

Represent and Solve More Take From Problems

Name ____________________

Review

You can use known numbers to find the unknown in a subtraction problem.

There are 11 apples. A family eats some. There are 7 left. How many does the family eat?

11 − ? = 7

You can use counters to solve. Start with 11 counters. Take away counters until 7 are left.

● ● ● ● ● ● ● ○ ○ ○ ○

11 − 4 = 7

The family eats 4 apples.

1. How can you make an equation? Use ? for the unknown. Use counters. Solve.

There are 12 crayons. Manny uses some. Now there are 6. How many does Manny use?

Equation:

● ● ● ● ● ●

_____ crayons

Represent and Solve More Take From Problems

Name ____________________

The table shows the prize tickets 3 friends earn.

Name	Mini-Golf	Ball Toss	Total
Mai	?	7	12
Ken	9	?	?
April	?	4	11

How can you make an equation? Solve.

1. How many prize tickets does Mai win playing mini-golf?

   ____

   ______ prize tickets

2. How many prize tickets does April win playing mini-golf?

   ____

   ______ prize tickets

3. Ken earns the most tickets. At least how many prize tickets does he win playing ball toss?

   ____

   ______ prize tickets

Represent and Solve Take Apart Problems

Name ____________________

Review

You can use a part and a difference to find an unknown total.

A bowl has green and red apples. 4 are green and 6 are red. How many apples are there?

$? - 4 = 6$

You can use connecting cubes to solve.
Start with 4 cubes. Add 6 cubes to get to 10.

$10 - 4 = 6$

There are 10 apples.

1. How can you make an equation? Use ? for the unknown. Draw connecting cubes. Solve.

There are some squirrels. 4 are brown and 3 are gray. How many squirrels are there?

Equation:

______ squirrels

Represent and Solve Take Apart Problems

Name

You decide to write word problems about two of your favorite people. How can you write a subtraction word problem with an unknown total about to match the picture? Solve the problem.

1.

Part	Part
4	13
Whole	
?	

2.

3. ● ● ● ● ● ● ○ ○ ○

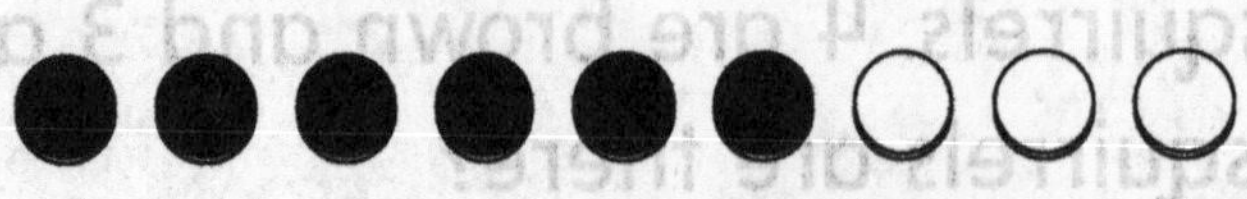

Represent and Solve More Take Apart Problems

Name ______________________________

Review

When both parts are unknown, many equations can match the problem.

Mel has 12 fish. Some are striped. The others are spotted. How many fish of each kind are there?

12 − ? = ?

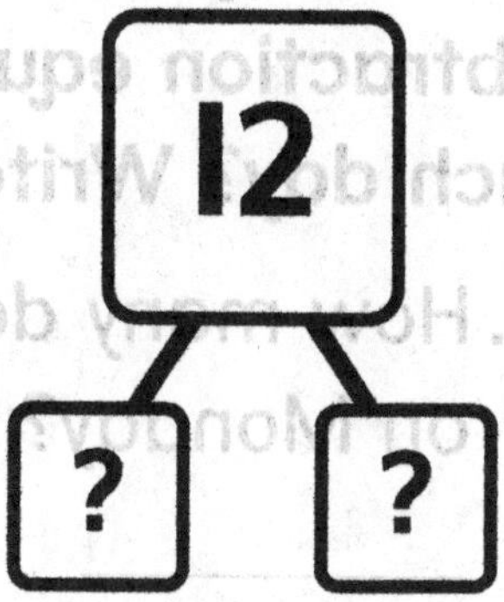

You can use a number bond to help you solve the problem.

These 3 equations match the problem:

12 − **5** = **7**   12 − **6** = **6**   and   12 − **8** = **4**.

5 striped and 7 spotted fish is a possible answer.

1. How can you make an equation? Use ? for the unknown. Use the number bond to solve. Write a possible answer.

A box holds 18 pencils. Some are red and the rest are blue. How many pencils are there?

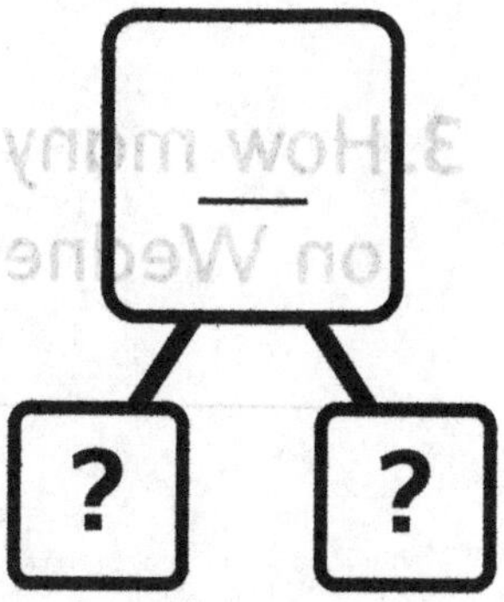

Equation:

____

Represent and Solve More Take Apart Problems

Name ________________________________

The table shows the total number of pets that go to the veterinarian. Each pet is a dog or cat.

Day	Pets
Monday	20
Tuesday	17
Wednesday	18

How can you make a subtraction equation to show how many pets go each day? Write 3 possible answers.

1. How many dogs and cats go to the veterinarian on Monday?

   ____

2. How many dogs and cats go to the veterinarian on Tuesday?

   ____

3. How many dogs and cats go to the veterinarian on Wednesday?

   ____

Solve Problems Involving Subtraction

Name ______________________________

Review

You can find an unknown part or difference in a word problem.

Jacob packs 14 socks for a trip. 10 are white and the rest are gray. How many socks are gray?

$14 - 10 = ?$

You can use connecting cubes to help. Start with 10. Add more until you get to the total, 14. You need 4 more.

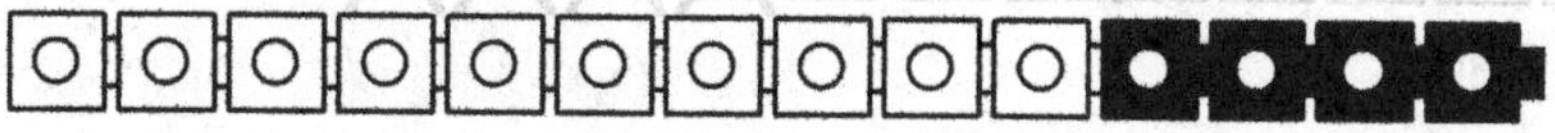

$14 - 10 = 4$, so Jacob packs 4 gray socks.

1. How can you make an equation to show the problem? Use ? for the unknown. Use connecting cubes to solve.

   A vase has 15 flowers. 9 are roses and the rest are daisies. How many daisies are there?

   Equation:

   ______ daisies

Solve Problems Involving Subtraction

Name ____________________

How can you write a subtraction word problem with an unknown part or difference to match the picture? Solve the problem.

1.

2.

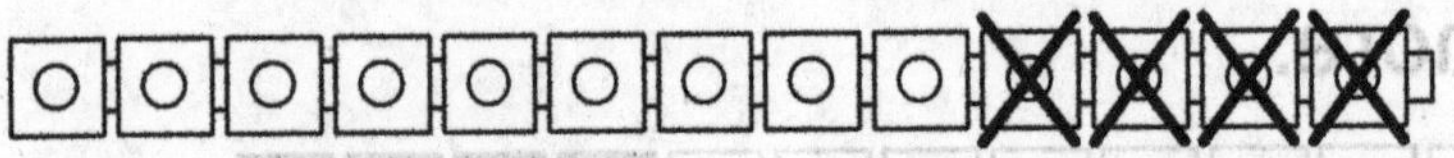

3.

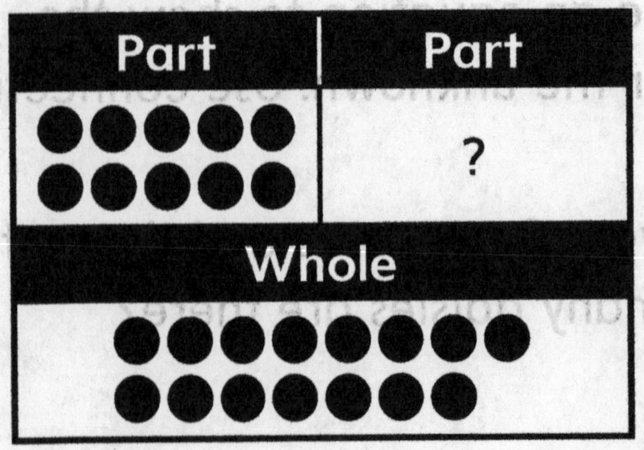

Lesson **8-6** • **Reinforce Understanding**

Solve More Problems Involving Subtraction

Name ______________________________

Review

You can find an unknown part or an unknown whole in word problems.

There are 12 superheroes in a movie. Some superheroes fly away. There are 3 superheroes left. How many superheroes fly away?

12 − **?** = **3**

You can use a number line to solve the problem.

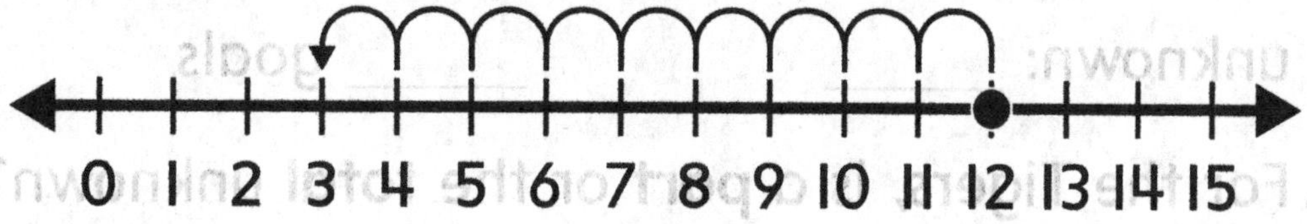

12 − **9** = 3 9 superheroes fly away.

1. How can you make an equation? Use ? for the unknown. Use the number line.

J.P. has 18 coins. He gives 10 coins away and keeps the rest. How many coins does he keep?

Equation:

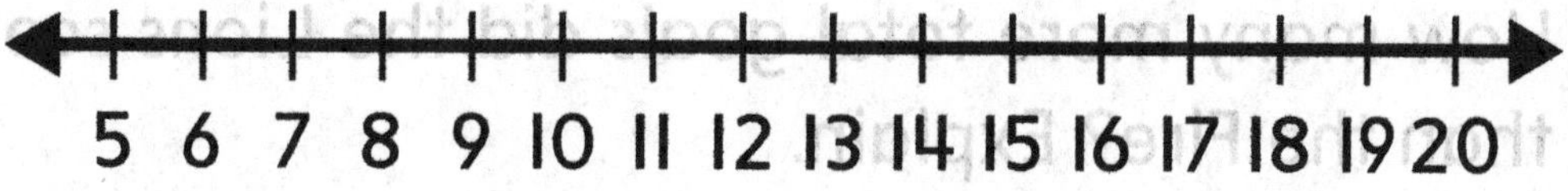

_______ coins

Solve More Problems Involving Subtraction

Name ______________________________

The standings show the number of goals scored by some soccer teams in 2 games.

Team	Game 1	Game 2	Total Goals
Fire		3	9
Tigers	8		12
Lions	7	6	

Is a part or the total unknown? Use the information to solve.

1. For the Fire, is a **part** or the **total** unknown?

   unknown: ________ ________ goals

2. For the Tigers, is a **part** or the **total** unknown?

   unknown: ________ ________ goals

3. For the Lions, is a **part** or the **total** unknown?

   unknown: ________ ________ goals

4. How many goals did all the teams score in their second games? Explain.

5. How many more total goals did the Lions score than the Fire? Explain.

Solve Problems Involving Addition and Subtraction

Name ______________________________

Review

You can find unknowns in word problems.

There are 12 cherries. Nick eats some. Now there are 7 cherries. How many cherries did Nick eat?

12 − ? = 7

You can use a number line to solve the problem.

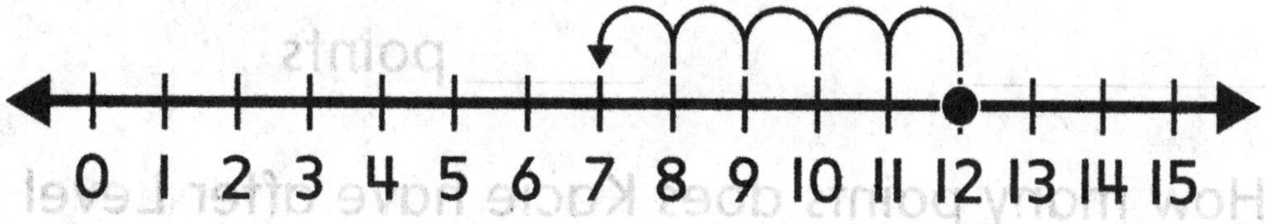

12 − **5** = 7

Nick ate 5 cherries.

1. How can you make an equation? Use ? for the unknown. Use the number line.

Mrs. Ling has 19 juice boxes in her refrigerator. She takes out 7 juice boxes. How many are left?

Equation:

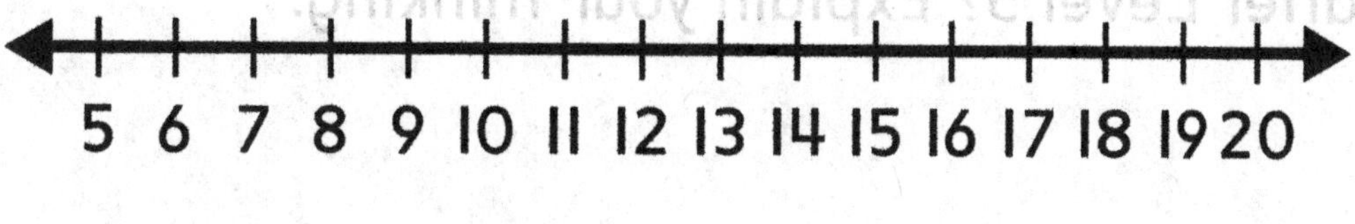

_______ juice boxes

Solve Problems Involving Addition and Subtraction

Name ______________________________

Kacie starts a video game with 10 points.

How can you make an equation? Solve.

Level	Change
1	Earn 6 points
2	Lose 7 points
3	Earn 8 points
4	Lose 9 points

1. How many points does Kacie have after Level 1?

   __________ ______ points

2. How many points does Kacie have after Level 2?

   __________ ______ points

3. How many points does Kacie have after Level 3?

   __________ ______ points

4. How many points does Kacie have after Level 4?

   __________ ______ points

5. How many points do you think Kacie will have after Level 5? Explain your thinking.

Use Mental Math to Find 10 More

Name ______________________________

Review

You can use a number line and patterns to find 10 more.

24 + 10 = ?

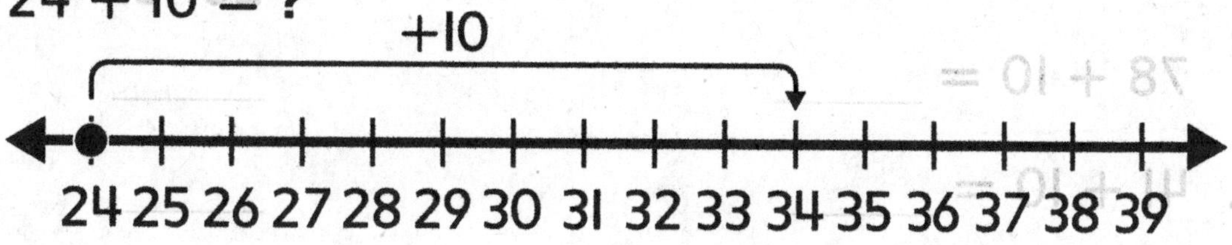

When you add 10, the tens digit goes up by 1 and the ones digit stays the same.

24 + 10 = **34**

What is the sum? Use the number line.

1. 56 + 10 = ______

2. 35 + 10 = ______

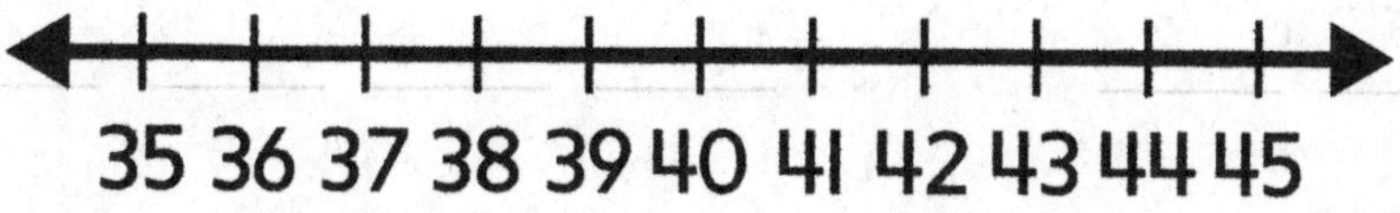

Use Mental Math to Find 10 More

Name ______________________________

What is the sum? Match the sum with the letter.

A	E	M	N	O	R	U	Y
53	42	17	76	51	60	37	88

Letter

1. 78 + 10 = ______ ______
2. 41 + 10 = ______ ______
3. 27 + 10 = ______ ______
4. 50 + 10 = ______ ______
5. 66 + 10 = ______ ______
6. 43 + 10 = ______ ______
7. 7 + 10 = ______ ______
8. 32 + 10 = ______ ______

Write the letters from top to bottom.

What belongs to you, but other people use it more than you do?

____ ____ ____ ____ ____ ____ ____ ____

Represent Adding Tens

Name ______________________________

Review

You can use a number chart to add tens.

27 + 40 = ?

1	2	3	4	5	6	7	8	9	10
11	12	13	14	15	16	17	18	19	20
21	22	23	24	25	26	27	28	29	30
31	32	33	34	35	36	37	38	39	40
41	42	43	44	45	46	47	48	49	50
51	52	53	54	55	56	57	58	59	60
61	62	63	64	65	66	67	68	69	70
71	72	73	74	75	76	77	78	79	80
81	82	83	84	85	86	87	88	89	90
91	92	93	94	95	96	97	98	99	100

27 + 40 = **67**

What is the sum?

1. 72 + 20 = ______ **2.** 34 + 30 = ______

3. 18 + 40 = ______ **4.** 41 + 20 = ______

5. 25 + 50 = ______ **6.** 56 + 30 = ______

Represent Adding Tens

Name ______________________________

A grocery store manager keeps track of what is in the store. Solve the problem.

Inventory
Carrots – 33 pounds
Lettuce – 18 cases
Apple – 27 crates
Potatoes – 56 bags

1. A truck delivers 50 pounds of carrots. How many pounds of carrots are there now?

   ______ pounds

2. A truck delivers 30 cases of lettuce. How many cases of lettuce are there now?

   ______ cases

3. A truck delivers 40 crates of apples. How many crates of apples are there now?

   ______ crates

4. There are 29 bags of potatoes on display. A clerk wants to put out 20 more bags. Are there enough bags in the store? Explain.

Represent Adding Tens and Ones

Name ________________________________

Review

You can add ones to a number.

32 + 5 = ?

Start at 32. Count on by ones 5 times.

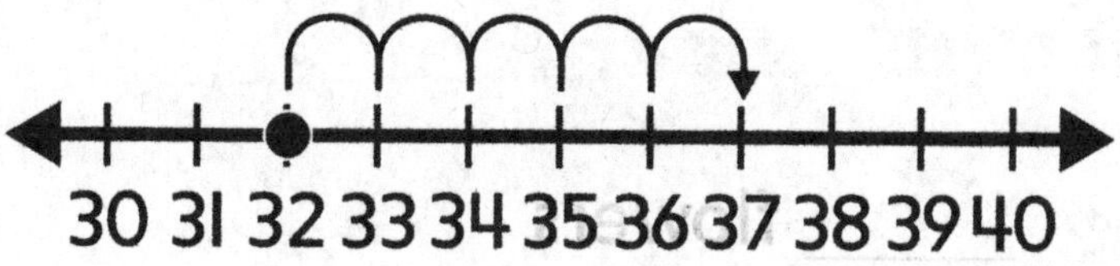

When you add ones to a number, you add the ones digits. The sum of the ones is less than 10, so the tens digit stays the same.

32 + 5 = **37**

What is the sum?

1. 74 + 4 = ______

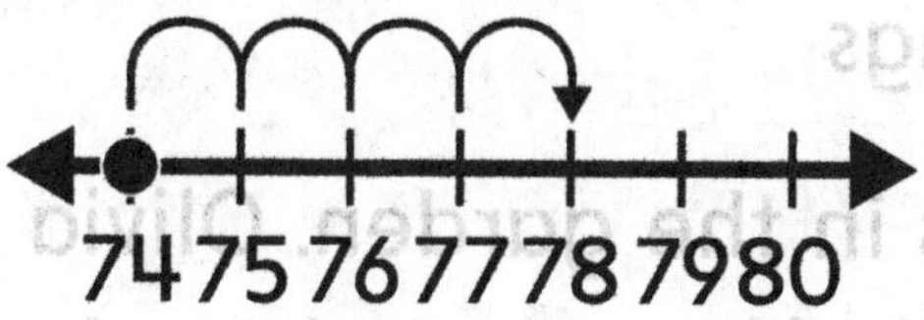

2. 25 + 3 = ______

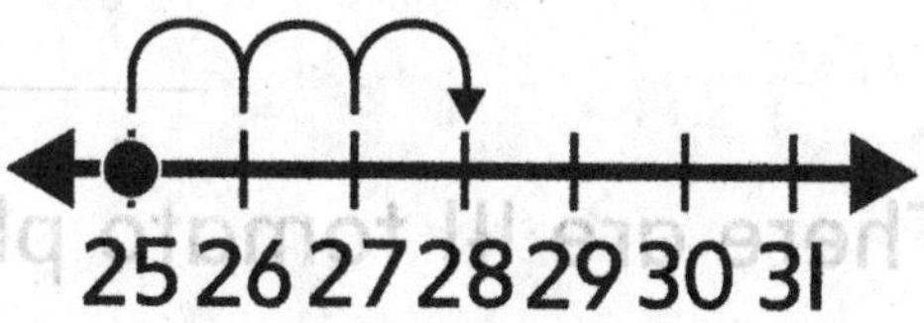

3. 26 + 3 = ______

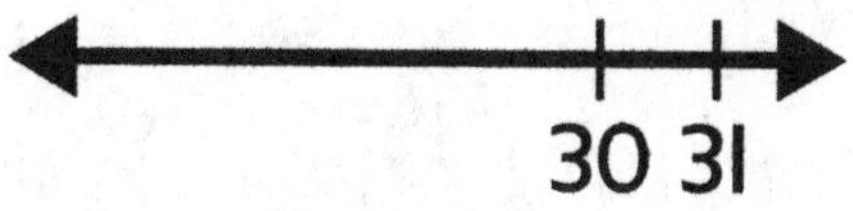

4. 41 + 5 = ______

5. 33 + 5 = ______

6. 51 + 6 = ______

Represent Adding Tens and Ones

Name ______________________________

Make a drawing to solve the problem.

1. Meghan plants 25 flowers. Then she plants 4 more flowers. How many flowers does Megan plant?

_____ flowers

2. Tom has 12 bags of mulch. He buys 6 more bags of mulch. How many bags of mulch does Tom have now?

_____ bags

3. There are 41 tomato plants in the garden. Olivia plants 7 more tomato plants. How many tomato plants are there now?

_____ tomato plants

Decompose Addends to Add

Name _____

Review

You can break apart one or both addends using a diagram.

$$22 + 34 = ?$$

22 + 34

20 + 2 30 + 4

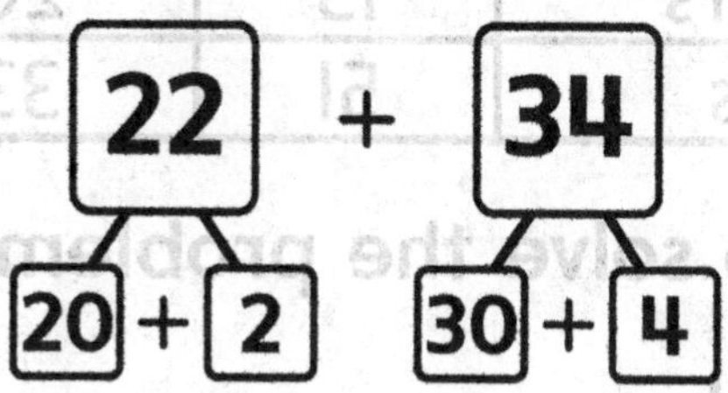

Add the tens: $20 + 30 = 50$

Add the ones: $2 + 4 = 6$

Add the tens and ones: $50 + 6 = 56$

So, $22 + 34 = 56$.

What is the sum?

1. $44 + 13 =$ _____

44 + 13

40 + 4 10 + 3

2. $25 + 34 =$ _____

25 + 34

☐ + ☐

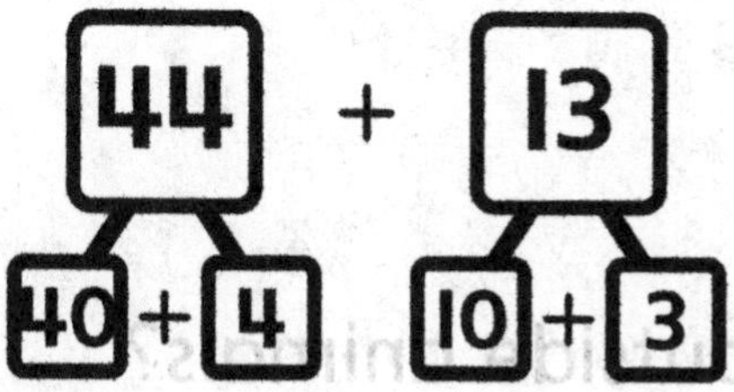

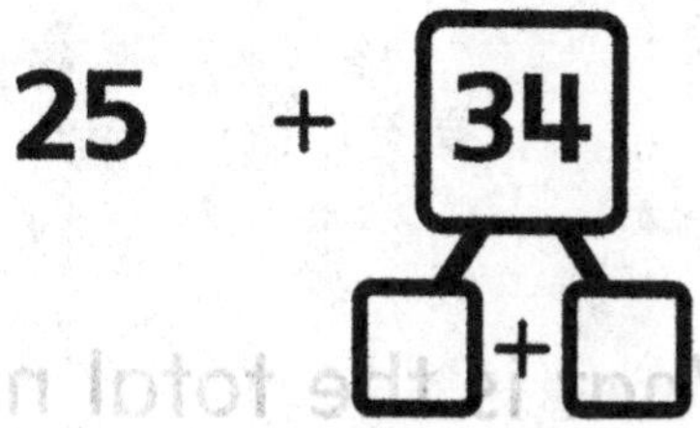

3. $12 + 54 =$ _____

4. $33 + 52 =$ _____

Decompose Addends to Add

Name ______________________________

The zoo has some animals that you can see both inside and outside.

Animal	Inside	Outside
Monkeys	12	42
Tigers	13	20
Birds	51	33

Use the table to solve the problems.
Show your work.

1. How many monkeys are there in all?

2. How many birds are there in all?

3. How many tigers are there in all?
Did you take apart an addend to find the sum?
Explain your thinking.

4. What is the total number of outside animals?
Explain your thinking.

Use an Open Number Line to Add within 100

Name ______________________

Review

You can use a number chart to help add 2-digit numbers.

31 + 25 = ?

Start at 31.
Decompose 25 as 20 + 5.
Move down 2 rows for the 2 tens.
Move across 5 columns for 5 ones.

21	22	23	24	25	26	27	28	29	30
31	32	33	34	35	36	37	38	39	40
41	42	43	44	45	46	47	48	49	50
51	52	53	54	55	56	57	58	59	60
61	62	63	64	65	66	67	68	69	70

So, 31 + 25 = **56**.

What is the sum? Use a number chart.

1. 42 + 17 = ______

31	32	33	34	35	36	37	38	39	40
41	42	43	44	45	46	47	48	49	50
51	52	53	54	55	56	57	58	59	60
61	62	63	64	65	66	67	68	69	70

2. 23 + 23 = ______

11	12	13	14	15	16	17	18	19	20
21	22	23	24	25	26	27	28	29	30
31	32	33	34	35	36	37	38	39	40
41	42	43	44	45	46	47	48	49	50
51	52	53	54	55	56	57	58	59	60

Use an Open Number Line to Add within 100

Name ______________________________

Some first and second grade students are asked to vote for their favorite field trip.

Field Trip	First Grade	Second Grade
Aquarium	34	42
Science Museum	13	40
Zoo	51	12

1. How many students vote for the aquarium?

   ______ students

2. Did more students vote for science museum or zoo? Explain your thinking.

3. Did more first graders or more second graders vote? Explain your thinking.

Decompose to Add on an Open Number Line

Name ______________________________

Review

You can decompose to add.

45 + 9 = ?

You can decompose 9 into 5 and 4.

45 + 5 = 50.

50 + 4 = 54.

What is the sum? Explain how you added.

1. 38 + 4 = ______

Break apart one addend: 4 = ______ + ______

Add: 38 + ______ = ______

Add: ______ + ______ = ______

2. 7 + 67 = ______

3. 79 + 6 = ______

Decompose to Add on an Open Number Line

Name ______________________________

A new bakery opens in town. Write and solve an equation. Explain how you used the strategy to solve.

1. Within the first hour of the grand opening, the bakery sold 34 loaves of white bread and 8 loaves of wheat bread. How many loaves did it sell altogether?

2. On Monday morning they sold 18 soft pretzels. During that afternoon they sold 7 soft pretzels. How many soft pretzels did the bakery sell on Monday?

3. During its first week the bakery sold 25 packages of blueberry muffins and 9 packages of bran muffins. How many packages of muffins did the bakery sell?

Regroup to Add

Name ________________

Review

You can use connecting cubes to add a 1-digit number to a 2-digit number.

28 + 6 = ?

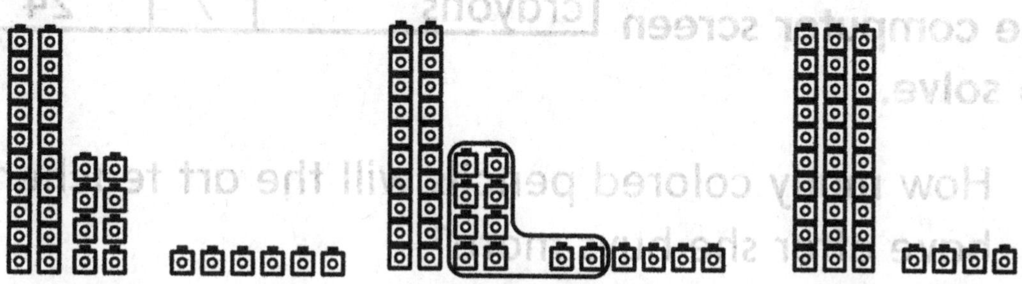

Model 28 + 6. Regroup 10 ones as 1 ten. The sum is 34.

What is the sum? Use the connecting cubes.

1. 35 + 8 = ______

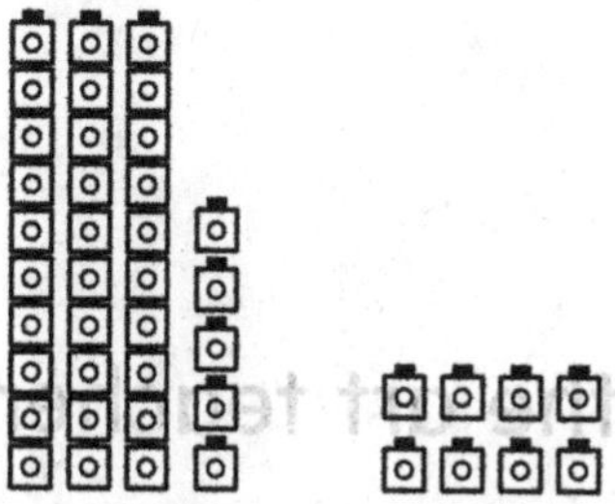

2. 7 + 24 = ______

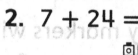

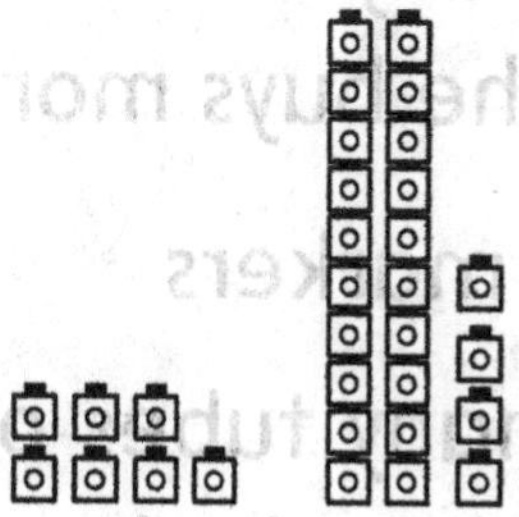

What is the sum?

3. 18 + 4 = ______

4. 5 + 59 = ______

5. 6 + 69 = ______

6. 48 + 8 = ______

Regroup to Add

Name ______________________________

An art teacher is making of list of the supplies she has and will buy. Use the information from the computer screen to solve.

	Has	Will Buy
colored pencils	14	9
paintbrushes	23	6
markers	27	8
tubes of paint	14	6
crayons	7	24

1. How many colored pencils will the art teacher have after she buys more?

______ colored pencils

2. How many paintbrushes will the art teacher have after she buys more?

______ paintbrushes

3. How many markers will the art teacher have after she buys more?

______ markers

4. How many tubes of paint will the art teacher have after she buys more?

______ tubes of paint

5. How many crayons will the art teacher have after she buys more?

______ crayons

Add 2-Digit Numbers

Name ______________________

Review

You can use connecting cubes to help you add 2-digit numbers.

34 + 18 = ?

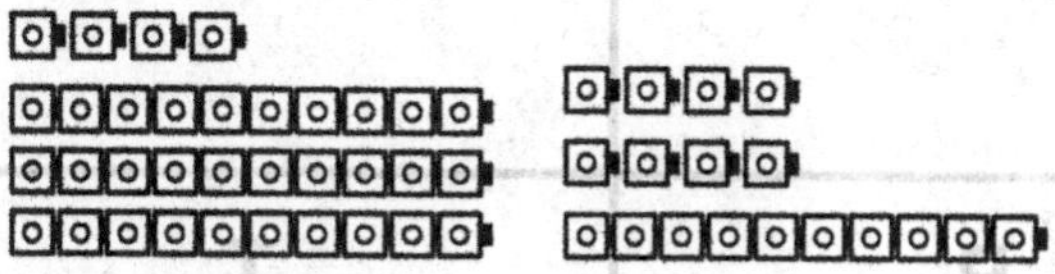	Model 34 + 18.
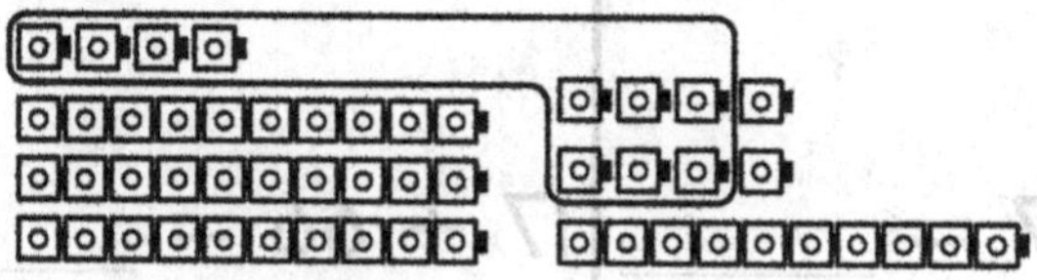	Regroup 10 ones as 1 ten.
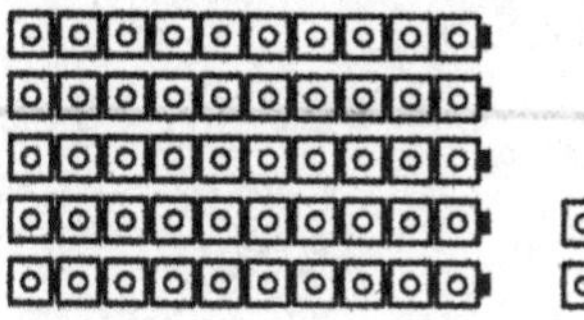	The sum is 52.

What is the sum? Use the connecting cubes.

1. 47 + 16 = ______

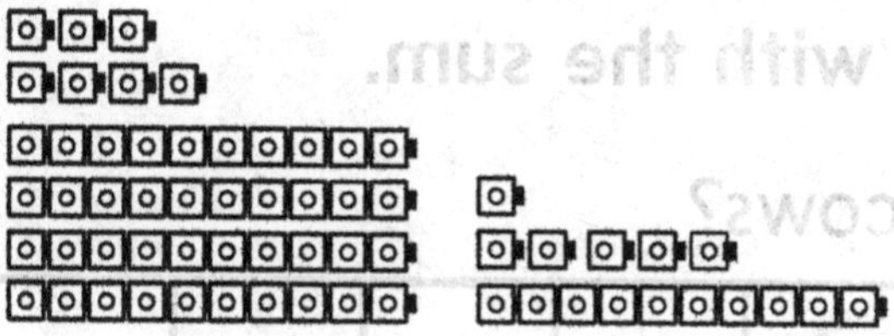

2. 27 + 64 = ______

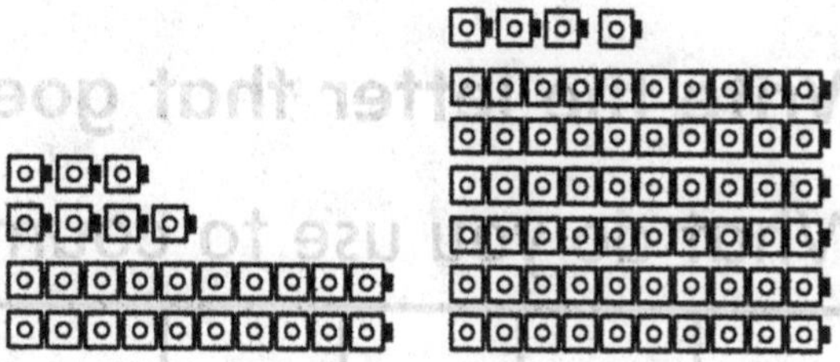

3. 54 + 39 = ______

4. 45 + 26 = ______

Add 2-Digit Numbers

Name ____________________

What is the sum?

A 38 + 55 = ____	**C** 39 + 17 = ____	**L** 16 + 16 = ____
O 59 + 18 = ____	**R** 18 + 17 = ____	**T** 17 + 66 = ____
U 45 + 36 = ____	**W** 43 + 19 = ____	

Write the letter that goes with the sum.

What do you use to count cows?

____	____	____	____	____	____	____	____	____	____
56	77	62	56	81	32	93	83	77	35

Represent and Solve Compare Problems

Name ______________________

Review

You can compare numbers to find the difference.

Timmy has 18 markers. Ana has 13 markers. How many more markers does Timmy have then Ana?

Use a number line. Start at 18. Count back to 13.

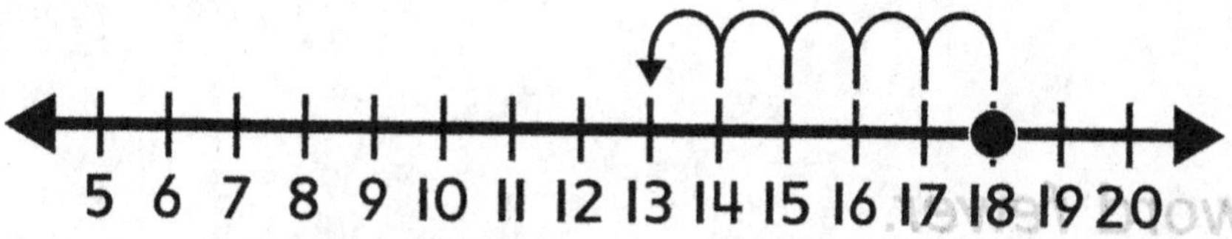

There are 5 jumps.

So, 18 − 13 = 5.

Timmy has 5 more markers than Ana.

1. Solve the problem. Use the number line.

Casey buys 19 game tokens. Jack buys 12 game tokens. How many more game tokens does Casey buy than Jack?

______ tokens

Represent and Solve Compare Problems

Name ______________________________

Write a word problem for the model. Use the compare word shown. Then solve.

1. Use the word *more*.

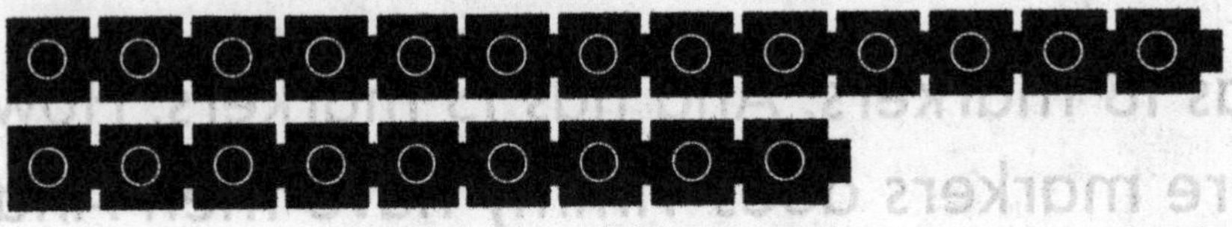

2. Use the word *fewer*.

3. Use the word *more*.

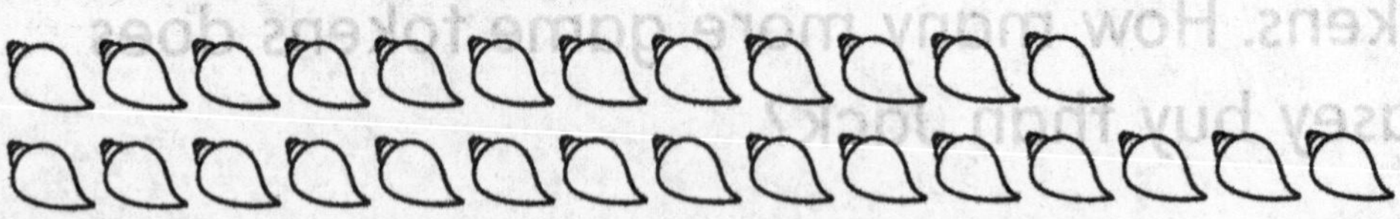

Represent and Solve Compare Problems Using Addition

Name ______________________________

Review

You can add to find the greater number.

Gavin has 5 more pencils than Lucy. Lucy has 7 pencils. How many pencils does Gavin have?

Use connecting cubes to find 5 more than 7.

Lucy
Gavin

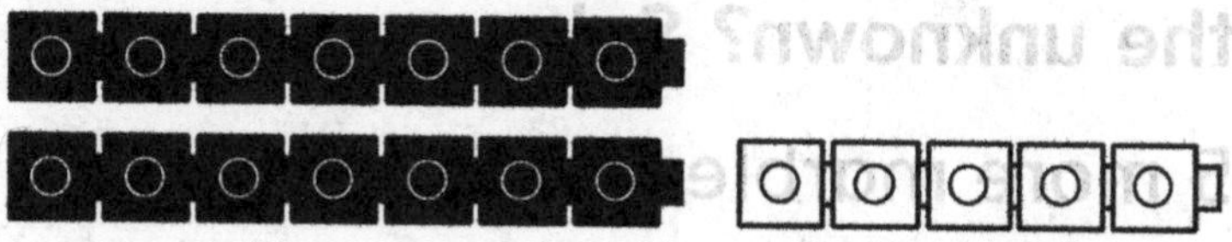

7 + 5 = 12, so Gavin has 12 pencils.

Solve. Show your thinking with connecting cubes.

1. Julia has 4 more books than Ben. Ben has 11 books. How many books does Julia have?

Ben
Julia

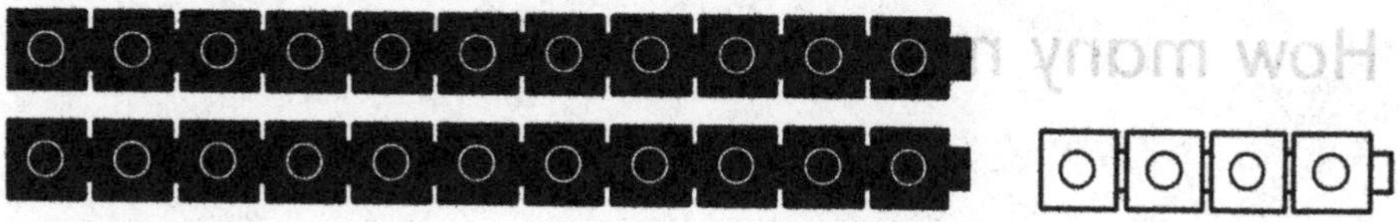

_______ books

2. Tri finds 3 fewer seashells than Edgar. Tri found 8 seashells. How many seashells does Edgar find?

_______ seashells

Represent and Solve Compare Problems Using Addition

Name ______________________________

The table shows how many marbles some students have.

Student	Marbles
Sawyer	9
Nick	14
Alfonso	12

Use the information in the table. How can you write an equation to represent the problem using ? for the unknown? Solve.

1. Jeff has 5 more marbles than Sawyer. How many marbles does Jeff have?

_____ marbles

2. Nick has 3 fewer marbles than Kylie. How many marbles does Kylie have?

_____ marbles

3. Caleb has 7 more marbles than Alfonso. How many marbles does Caleb have?

_____ marbles

Lesson 10-3 • Reinforce Understanding

Represent and Solve More Compare Problems

Name ____________________

Review

You can find the lesser number by using a number path.

Craig buys 3 more apples than Dana. Craig buys 7 apples. How many apples does Dana buy?

Start at 7. Count back 3. You end at 4.

1	2	3	4	5	6	7	8	9	10

7 − 3 = 4, so Dana buys 4 apples.

Solve. Use the number path.

1. Tom has 2 more blocks than Evan. Tom has 9 blocks. How many blocks does Evan have?

1	2	3	4	5	6	7	8	9	10

______ blocks

2. Lin has 4 fewer stickers than Dan. Dan has 10 stickers. How many stickers does Lin have?

1	2	3	4	5	6	7	8	9	10

______ stickers

Represent and Solve More Compare Problems

Name ____________________

Match the equation to another equation that can be used to show the same compare problem.

1. $12 - 7 = ?$	$11 - 6 = ?$
2. $5 + ? = 11$	$5 + ? = 12$
3. $6 + ? = 11$	$7 + ? = 12$
4. $12 - 5 = ?$	$11 - 5 = ?$

5. Make a compare word problem using one of the equations above. Then solve. Show your thinking with a drawing.

Solve Compare Problems Using Addition and Subtraction

Name ______________________________

Review

You can use tens and ones blocks to compare

There are 4 fewer girls than boys.
There are 13 girls.

How many boys are there?

Show 13 girls with 1 tens block and 3 ones blocks. Add 4 more ones blocks to show the boys.

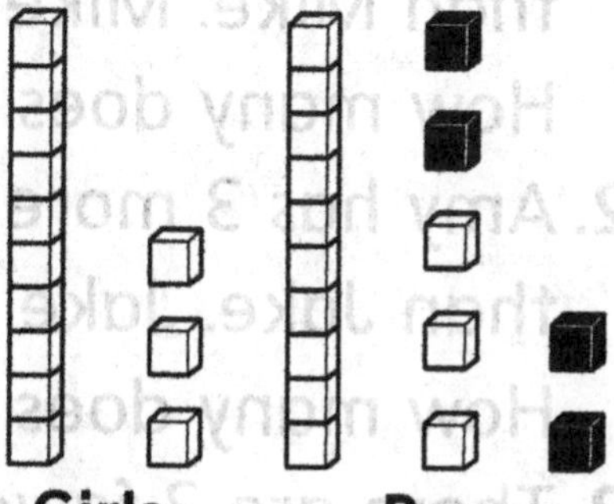

13 + 4 = 17

There are 17 boys.

1. Solve. Use tens and ones blocks to solve.

Jenny mails 4 fewer letters than Pablo. Jenny mails 12 letters. How many letters does Pablo mail?

______ letters

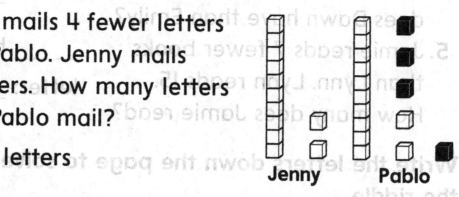

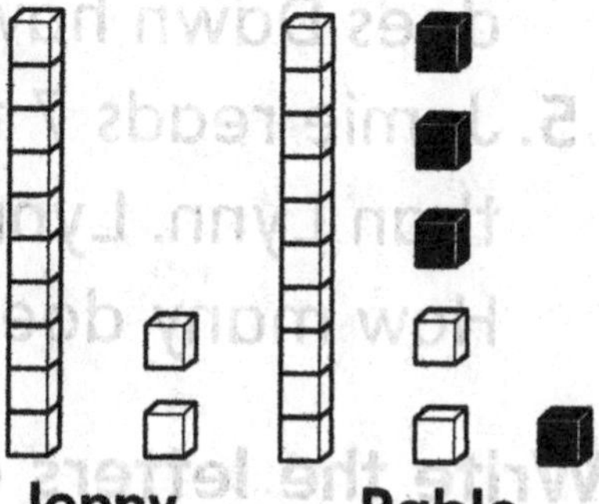

Solve Compare Problems Using Addition and Subtraction

Name ____________________

Solve. Then write the letter from the table.

Answer	E	L	O	T	W
Letter	7	8	11	9	5

1. Ling has 4 fewer balloons than Mike. Mike has 13. How many does Ling have?
   ______ balloons
   letter: ______
2. Amy has 3 more pencils than Jake. Jake has 8. How many does Amy have?
   ______ pencils
   letter: ______
3. There are 3 fewer balls than cones. There are 8 cones. How many balls are there?
   ______ balls
   letter: ______
4. Dawn has 8 fish. Emily has 15 fish. How many fewer fish does Dawn have than Emily?
   ______ fish
   letter: ______
5. Jamie reads 7 fewer books than Lynn. Lynn reads 15. How many does Jamie read?
   ______ books
   letter: ______

Write the letters down the page to solve the riddle.

Riddle: What gets wetter the more it dries?

Answer: A ______!

Use Mental Math to Find 10 Less

Name ______________________

Review

You can use a place-value chart to find 10 less.

45 − 10 = ?

Use a place-value chart. Think about the ones and tens in 45.

tens	ones
4	5

When you subtract 10, there will be one less ten.

45 − 10 = 35

tens	ones
~~4~~ 3	5

What is the difference? Use the place-value chart.

1. 39 − 10 = _____

tens	ones
~~3~~ 2	9

2. 81 − 10 = _____

tens	ones
8	

What is the difference?

3. 47 − 10 = _____

4. 92 − 10 = _____

Use Mental Math to Find 10 Less

Name ______________________________

The table shows the prizes at a carnival game at the beginning of the day.

Item	Number
Stuffed Animal	72
Beach Ball	56
Kite	69

Use the information in the table. How can you find 10 less? Solve.

1. People won 10 stuffed animals today. How many stuffed animals are left?

   ______ stuffed animals

2. People won 10 beach balls today. How many beach balls are left?

   ______ beach balls

3. People won 10 kites today. How many kites are left?

   ______ kites

4. Explain how you can use mental math to subtract 10 from a 2-digit number.

Representing Subtracting Tens

Name ____________________

Review

You can show subtracting tens different ways.

50 − 20 = ?

Show 50 with 5 blocks of 10. Cross out 2 blocks of 10.

There are 3 blocks of 10 left. So, 50 − 20 = 30.

What is the difference? Use the blocks of 10.

1. 70 − 30 = ______

2. 60 − 40 = ______

What is the difference?

3. 50 − 20 = ______   **4.** 80 − 60 = ______

Representing Subtracting Tens

Name ______________________________

Write a subtraction word problem for the model shown. Solve your word problem.

1.

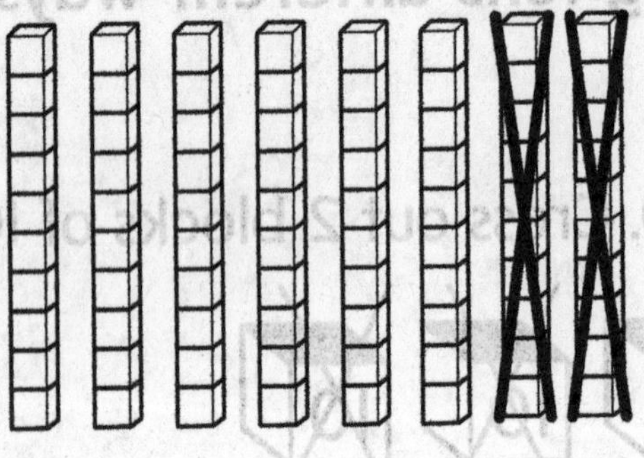

2.

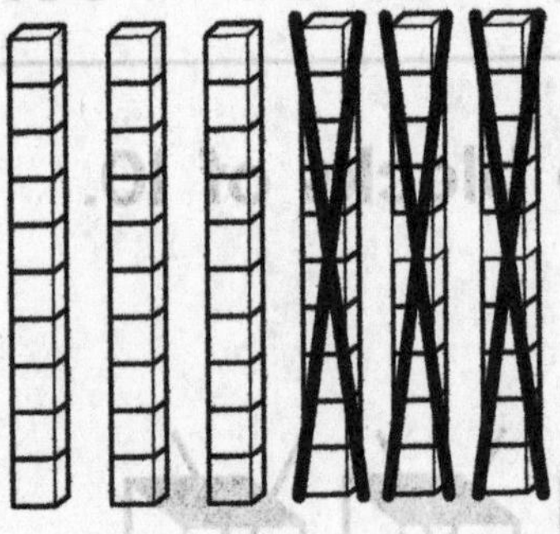

3.

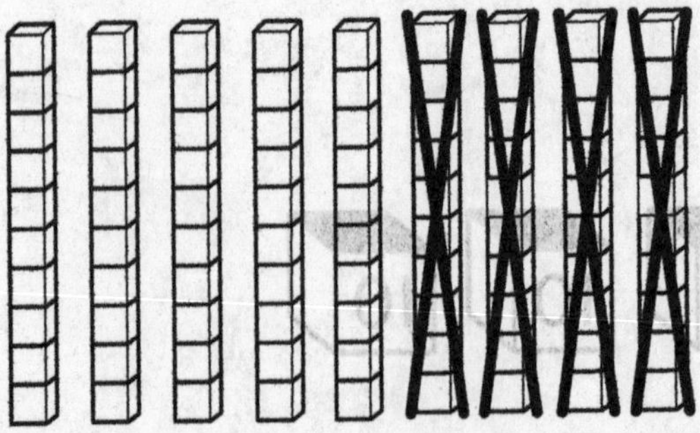

Subtract Tens

Name ______________________________

Review

You can use a number chart to subtract tens from tens.

70 − 30 = ?

Start at 70.

Move up 3 rows.

70 − 30 = 40

21	22	23	24	25	26	27	28	29	30
31	32	33	34	35	36	37	38	39	40
41	42	43	44	45	46	47	48	49	50
51	52	53	54	55	56	57	58	59	60
61	62	63	64	65	66	67	68	69	70

What is the difference? Use the number chart.

1. 60 − 20 = ? ______

21	22	23	24	25	26	27	28	29	30
31	32	33	34	35	36	37	38	39	40
41	42	43	44	45	46	47	48	49	50
51	52	53	54	55	56	57	58	59	60
61	62	63	64	65	66	67	68	69	70

What is the difference?

2. 90 − 60 = ? ______

3. 80 − 70 = ? ______

4. 70 − 20 = ? ______

5. 30 − 20 = ? ______

Subtract Tens

Name ______________________________

What is the unknown number in the subtraction equation?

1.

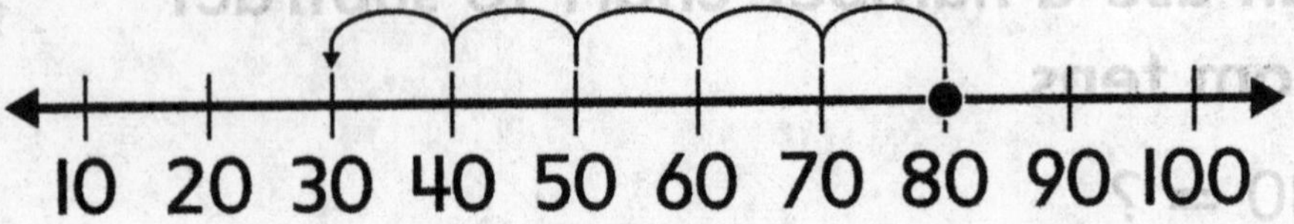

80 − 50 = _____

2.

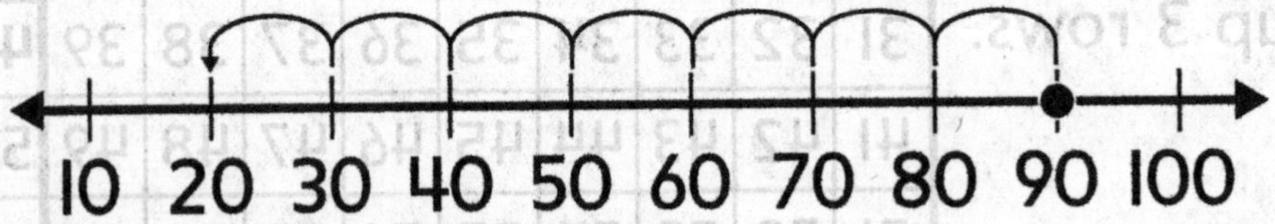

90 − _____ = 20

3.

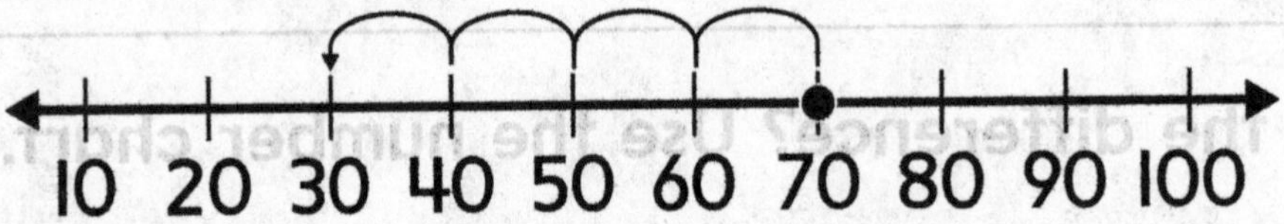

_____ − 40 = 30

4.

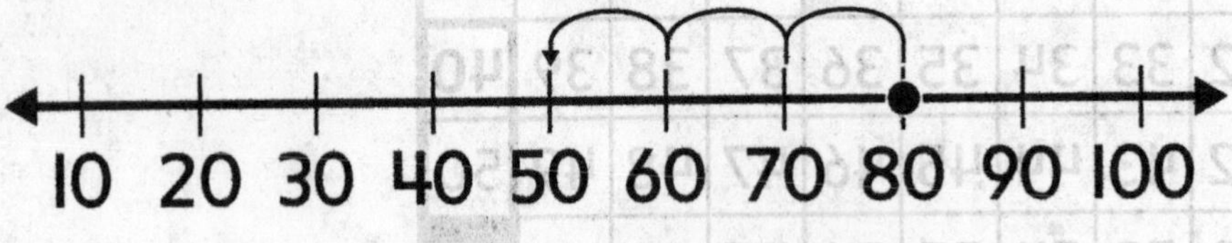

80 − _____ = 50

5.

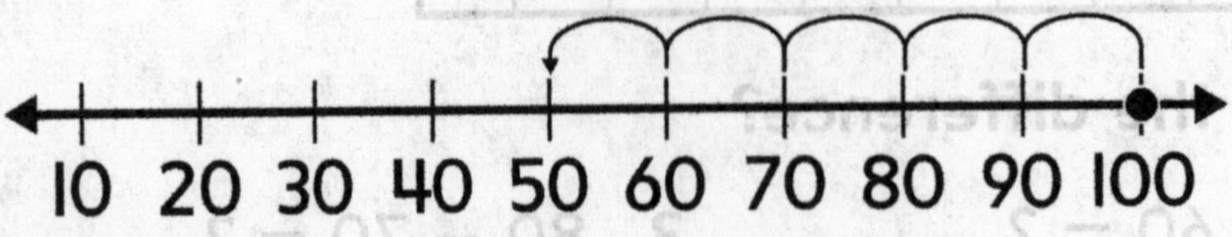

_____ − 50 = 50

Use Addition to Subtract Tens

Name ______________________________

Review

You can use addition to subtract tens.

80 − 30 = ?

Make an unknown addend equation.

30 + ? = 80

Use a number line to solve.

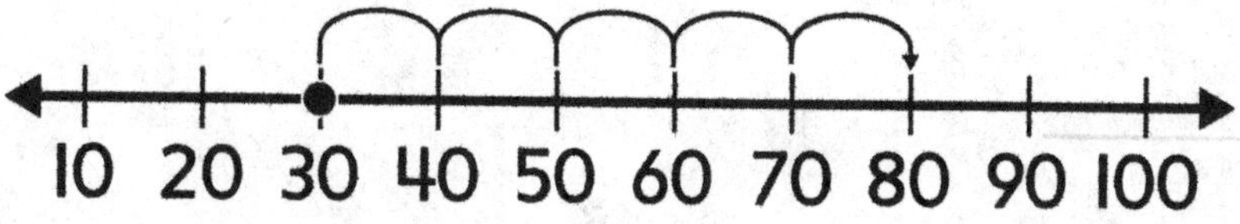

30 + **50** = 80, so 80 − 30 = **50**.

What addition equation can help you subtract? Show your thinking on a number line.

1. 70 − 40 = ______

Equation: ______________________________

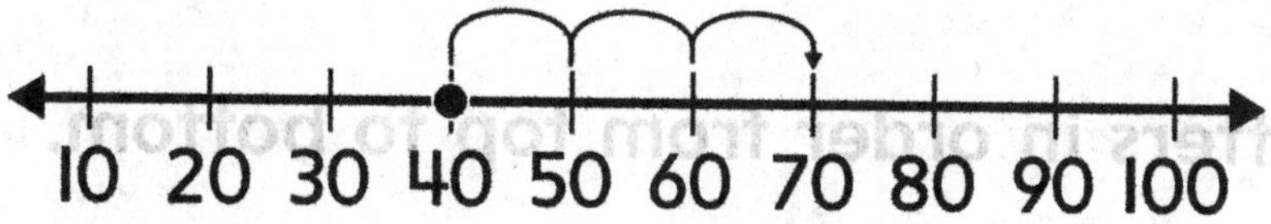

2. 90 − 30 = ______

Equation: ______________________________

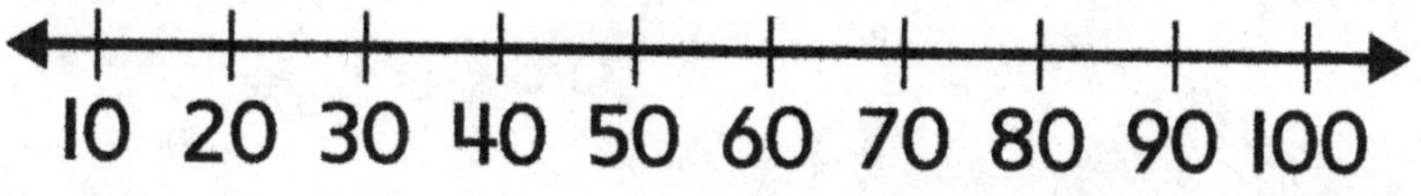

Use Addition to Subtract Tens

Name ______________________________

What is the difference? Write the letter from the table for the difference.

Difference	0	10	20	30	40	50	60	70	80	90
Letter	M	P	C	A	D	E	L	I	T	S

1. 70 − 60 = ______

letter: ______

2. 40 − 10 = ______

letter: ______

3. 90 − 30 = ______

letter: ______

4. 50 − 50 = ______

letter: ______

Write the letters in order from top to bottom.

Riddle: What kind of tree can you grow in your hand?

Answer: A ______ tree!

Explain Subtraction Strategies

Name ____________________

Review

You can use different strategies to subtract tens.

Explain the strategy used to subtract 90 − 30.

Show 9 tens.

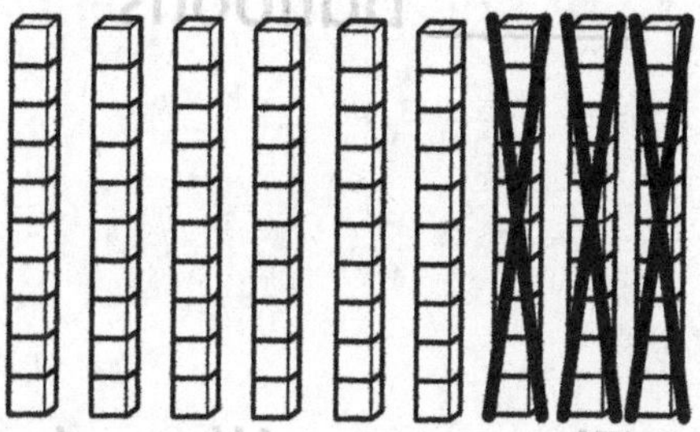

Cross out 3 tens.

There are 6 tens left.

So, 90 − 30 = **60**.

What is the difference? Explain the strategy used. Complete the sentences.

1. 60 − 10 = _____

11	12	13	14	15	16	17	18	19	20
21	22	23	24	25	26	27	28	29	30
31	32	33	34	35	36	37	38	39	40
41	42	43	44	45	46	47	48	49	50
51	52	53	54	55	56	57	58	59	60
61	62	63	64	65	66	67	68	69	70

Start at _____.

Move up _____ row.

You end at _____.

2. 40 − 40 = _____

40 + ? = 40 and

4 + _____ = 4, so

40 − 40 = _____.

Explain Subtraction Strategies

Name ______________________________

Solve the problem. Explain the strategy you used.

1. A store has 79 balloons. The store sells 30 balloons for a party. How many balloons are left?

_____ balloons

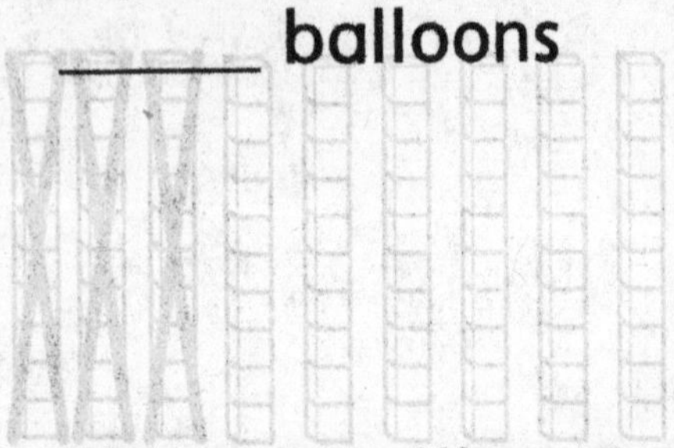

2. There are 64 customers in the store. 20 customers leave. How many customers are left?

_____ customers

3. Janie has 50 price-tag stickers. She uses 20 stickers. How many stickers are left?

_____ stickers

Compare and Order Lengths

Name ______________________________

Review

You can compare and order objects by length.

You can line up the ends of objects to compare lengths.

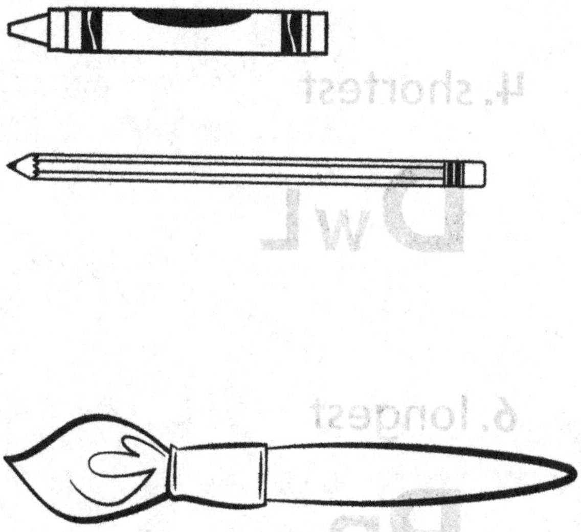

The crayon is the **shortest**.

The pencil is **shorter** than the paintbrush and **longer** than the crayon.

The paintbrush is the **longest**.

How can you order the objects by length? Write *longest* or *shortest* next to the correct objects.

1.

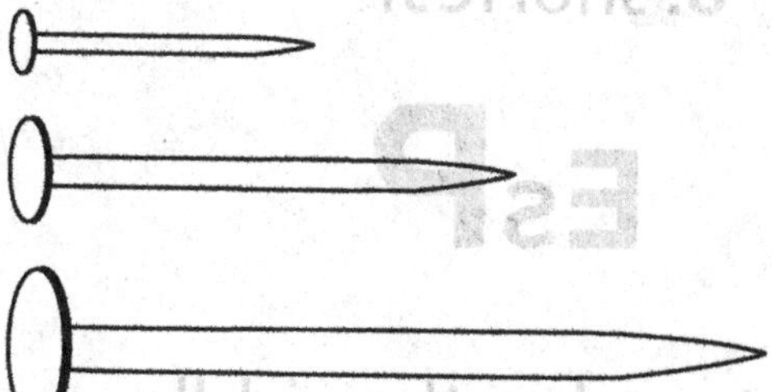

____

____

____

2.

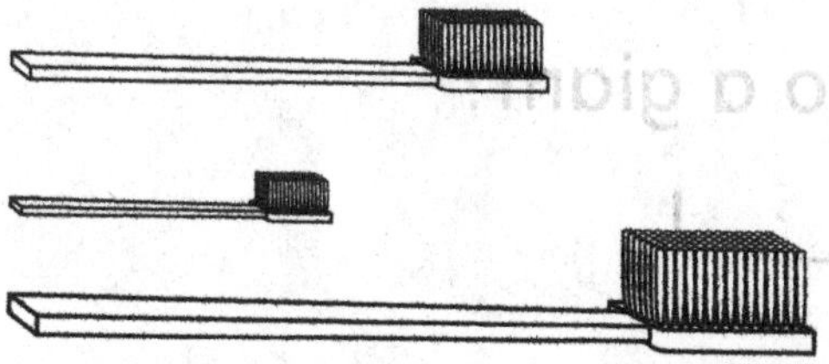

____

____

____

Compare and Order Lengths

Name ____________________

Circle the *shortest* or *longest* letter in the group.

1. longest

T A B

2. shortest

I C E

3. shortest

A P G

4. shortest

D W L

5. longest

E U O

6. longest

R B I

7. longest

T G D

8. shortest

E S P

Write the letters in order to solve the riddle.

Riddle: How do you talk to a giant?

Answer: Use _____ ________!

More Ways to Compare Lengths

Name ____________________

Review

You can compare the lengths of two objects using a third object.

You can use the length of the spoon to compare the lengths of the paper clip and rope.

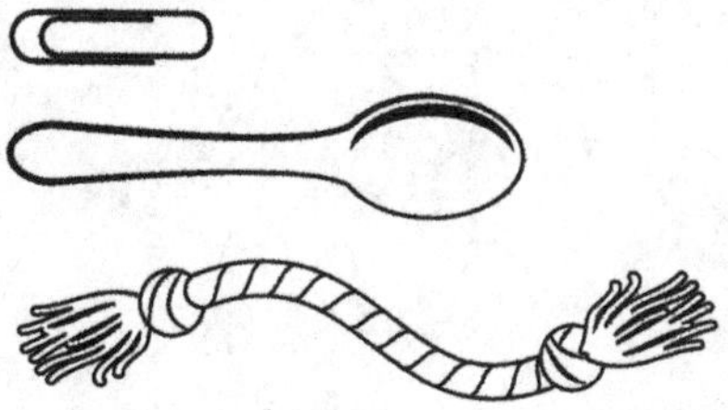

The paper clip is shorter than the spoon.

The rope is longer than the spoon.

So, the rope is longer than the paper clip.

1. Is the dropper shorter or longer than the toothbrush? Explain your thinking.

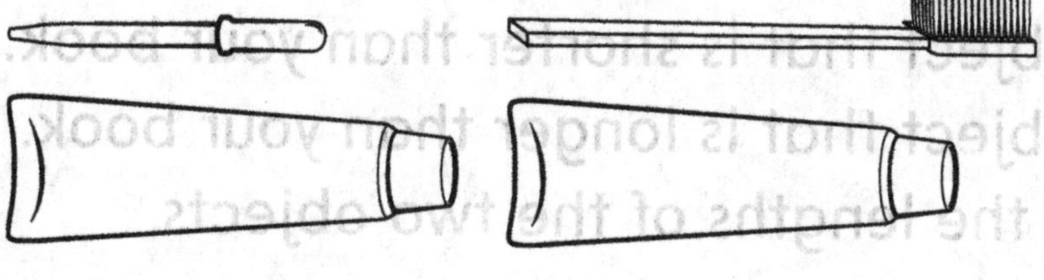

The dropper is __________ than the toothbrush.

More Ways to Compare Lengths

Name ______________________________

Find two objects in your classroom or school. Use the given object to compare lengths. Explain your thinking.

1. Find an object that is shorter than your hand.
   Find an object that is longer than your hand.
   Compare the lengths of the two objects.

2. Find an object that is longer than your shoe.
   Find an object that is shorter than your shoe.
   Compare the lengths of the two objects.

3. Find an object that is shorter than your book.
   Find an object that is longer than your book.
   Compare the lengths of the two objects.

Strategies to Measure Lengths

Name ______________________________

Review

You can lay units end to end to measure length.

How can you tell the length of the marker?

Choose a unit that is smaller than the marker such as a paper clip.

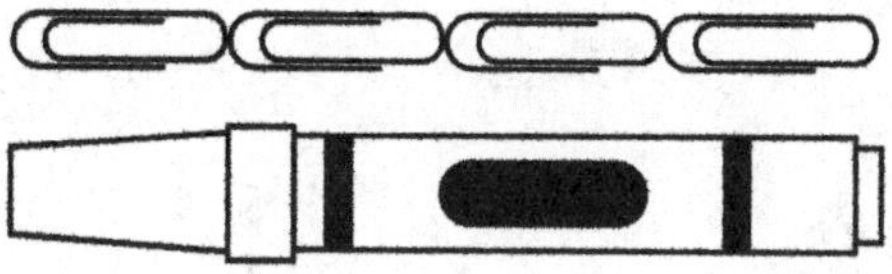

Lay paper clips end to end to match the length of the marker. Then count the paper clips.

The marker is 4 paper clips long.

1. How many connecting cubes long is the feather?

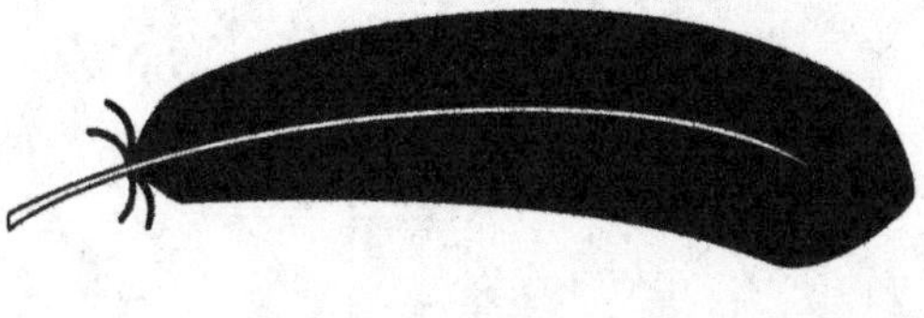

______ connecting cubes

Strategies to Measure Lengths

Name ______________________________

Choose a unit that can be used to measure the length of each object around your school. Explain your thinking. Then use the unit to measure the length.

1. the length of your pencil

2. the length of a book

3. the length of a stapler

4. the length of a lunch tray

More Strategies to Measure Lengths

Name ____________________

Review

You can use different units to measure the same object.

How can you tell the length of the toy train 2 different ways?

Choose 2 units.

Lay pennies and blocks end to end to match the length of the train. Then count the units.

The train is 8 pennies long.
The train is 6 blocks long.

1. How many connecting cubes and crayons long is the toy bowling pen?

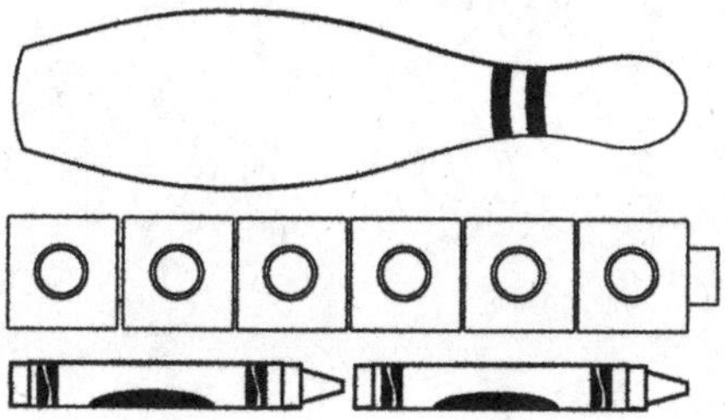

______ connecting cubes

______ crayons

Lesson 12-4 • Extend Thinking

More Strategies to Measure Lengths

Name ______________________________

Choose two units that can be used to measure the length of the object. Predict how many of each unit you will use. Explain your thinking. Then use the units to measure the length.

1. the length of your desk

2. the length of an eraser

3. the length of a glue stick

Tell Time to the Hour

Name ______________________________

Review

You can use analog and digital clocks to tell time.

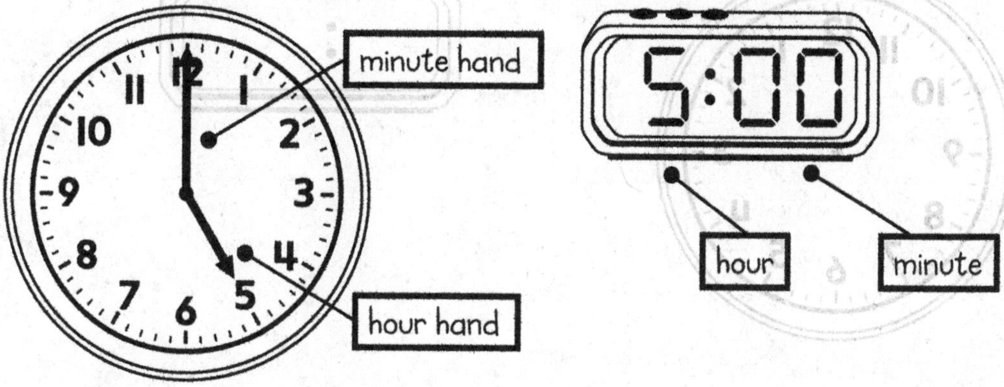

The short hand is the **hour hand.**

The long hand is the **minute hand.**

The first number shows 5 hours.

The second number shows 0 minutes.

The clocks show that the time is 5 o'clock.

What time is it? Write the time.

1.

_______:_______

2.

_______:_______

Tell Time to the Hour

Name ______________________________

Draw or write the time on the clocks for the event.

1. practice at 5:00

2. get up at 7 o'clock

3. school at 8 o'clock

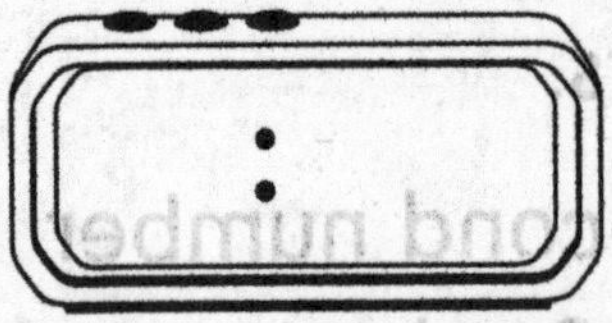

4. eat dinner at 6 o'clock

5. dentist at 2:00

6. recess at 11 o'clock

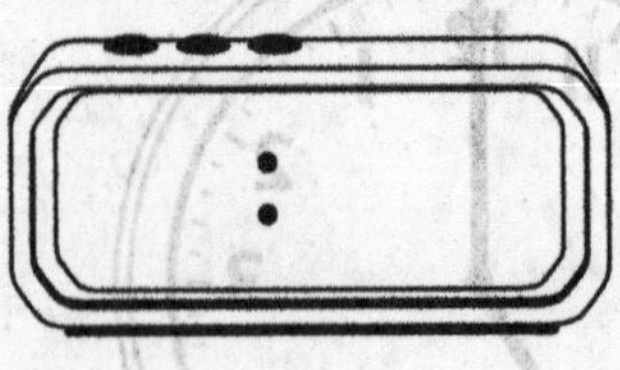

Tell Time to the Half Hour

Name ______________________________

Review

You can use clocks to tell time to the half hour.

The hour hand is between 8 and 9. The hour is 8.

The minute hand points to 6. The minutes are 30.

The first number shows 8 hours.

The second number shows 30 minutes.

The time is 8:30 or half past 8.

1. The hour hand is between 4 and 5. The hour is ______.
   The minute hand points to 6.
   The minutes are ______.
   It is ______ : ______.

Tell Time to the Half Hour

Name ____________________

The Owen's family schedule for the day is shown. Draw or write the time on the clocks for the event.

- **Kristie: piano lesson at 4:30**
- **Family dinner: half past 6**
- **Evan: math club at 3:30**
- **Daniel: practice at half past 5**

1. Kristie's piano lesson

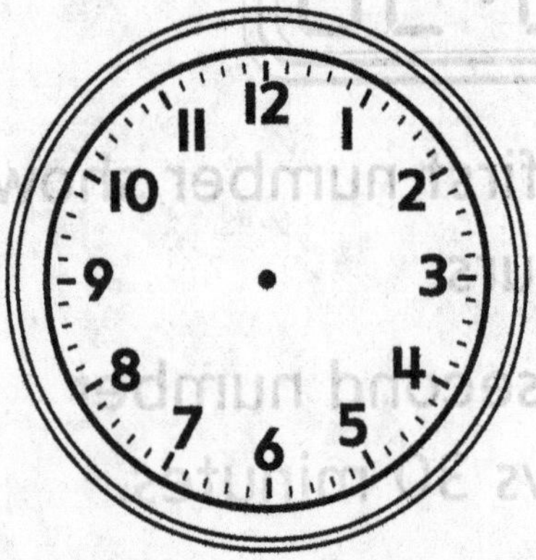

2. Family dinner

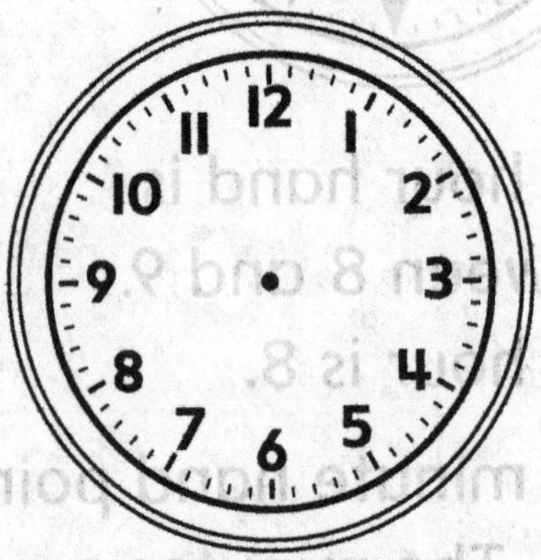

3. Evan's math club

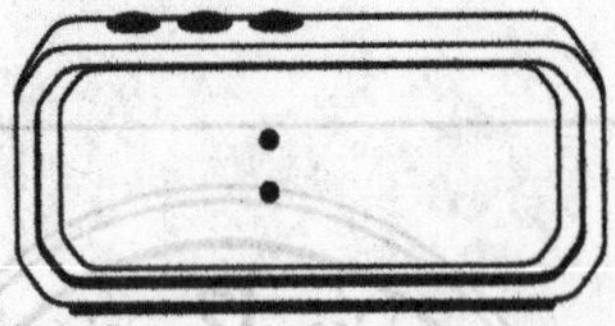

4. Daniel's practice

5. How do you draw or write the time to the half hour on an analog clock?

Organize Data

Name ____________________

Review

You can sort objects into categories in a chart.

How can you sort these objects by size?
Write the object names in the chart.

barn

paper clip

chair

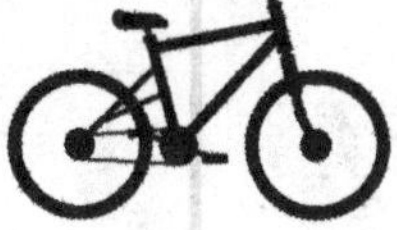
bike

coin

Small	Medium	Large
paper clip	chair	barn
coin	bike	

1. How can you sort these objects by shape?
Write the object names in the chart.

orange

puzzle

juice

block

blueberry

peanut butter

Sphere	Cube	Cylinder

Organize Data

Name ______________________________

Find groups of 5 objects in your school that can be sorted. Write the names in the chart.

1. Sort by shape.

Rectangle	Circle	Triangle

2. Sort by size.

Small	Medium	Large

3. Choose your own categories.

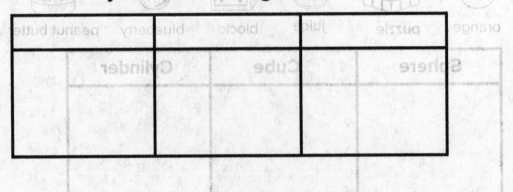

Represent Data

Name ____________________

Review

You can show how many of each kind of object in a tally chart.

Students picked their favorite instrument.

Make a tally mark for each instrument.

Then count the marks and write the total.

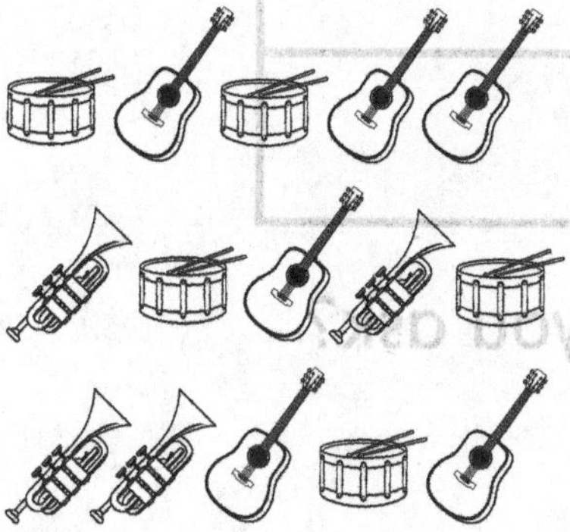

Favorite Instrument		
Instrument	**Tally**	**Total**
	𝍸	5
	𝍸 𝍷	6
	𝍷𝍷𝍷𝍷	4

1. Complete the tally chart.

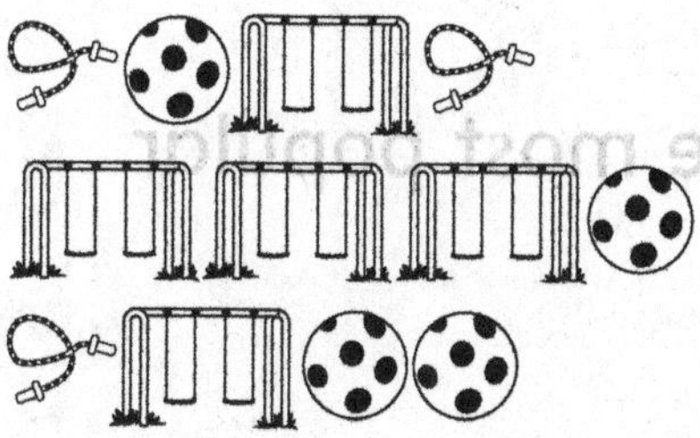

Favorite Recess Activity		
Activity	**Tally**	**Total**
	𝍷𝍷𝍷	3
	𝍷𝍷𝍷𝍷	___
		___

Represent Data

Name ______________________________

Ask 15–20 students at your school if they like music, reading, or math the best. Use the data to complete the chart.

Favorite School Subject		
School Subject	**Tally**	**Total**

123		

1. How many total students did you ask?

   ________ students

2. Of the students you asked, is music, reading, or math the most popular?

   ____

3. How many students chose the most popular school subject?

   ________ students

Interpret Data

Name ____

Review

You can use a picture graph to interpret data.

How many students picked each sport as their favorite?

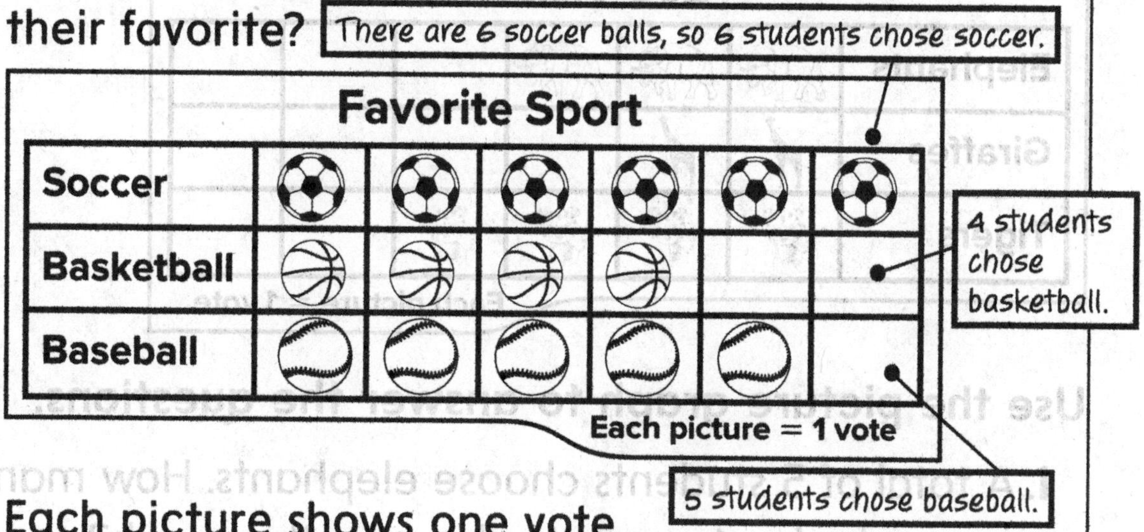

Each picture shows one vote.

Use the picture graph to answer the questions.

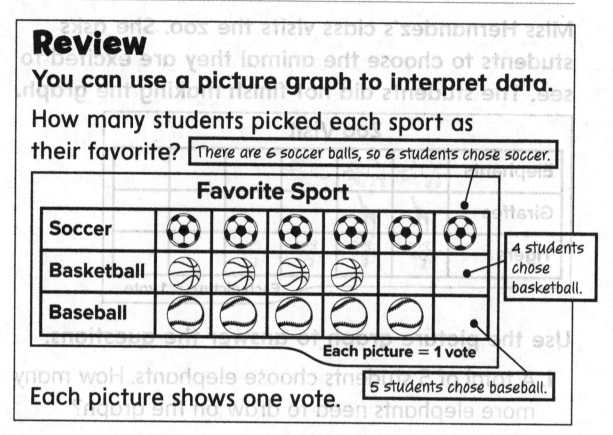

1. How many students chose dogs? _____ students
2. How many students chose cats? _____ students
3. How many students chose fish? _____ students

Lesson 12-9 • Extend Thinking

Interpret Data

Name ______________________________

Miss Hernandez's class visits the zoo. She asks students to choose the animal they are excited to see. The students did not finish making the graph.

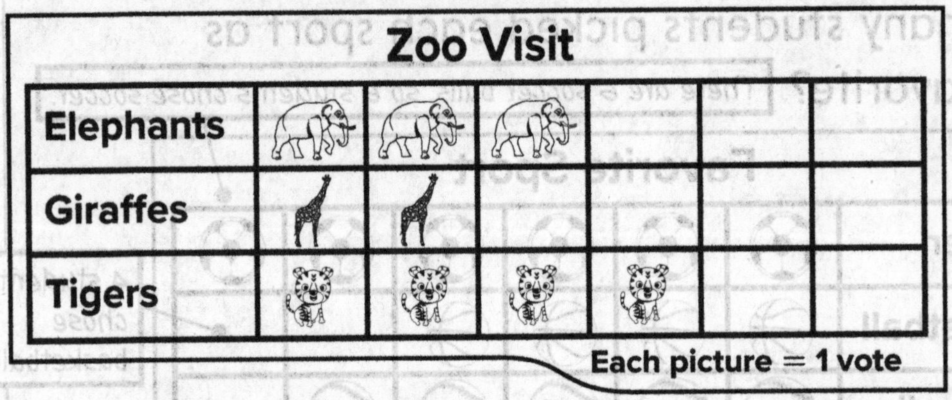

Use the picture graph to answer the questions.

1. A total of 5 students choose elephants. How many more elephants need to draw on the graph?

_______ elephants

2. Giraffes get 1 more vote than elephants. How many more giraffes need to draw on the graph?

_______ giraffes

3. Which zoo animal are more students excited to see than any other?

____

4. How many students does Miss Hernandez ask?

_______ students

Solve Problems Involving Data

Name ______________________________

Review

You can tell *how many more* and *how many less.*

Molly asks students to pick their favorite fruit. How many more students pick grapes than apples?

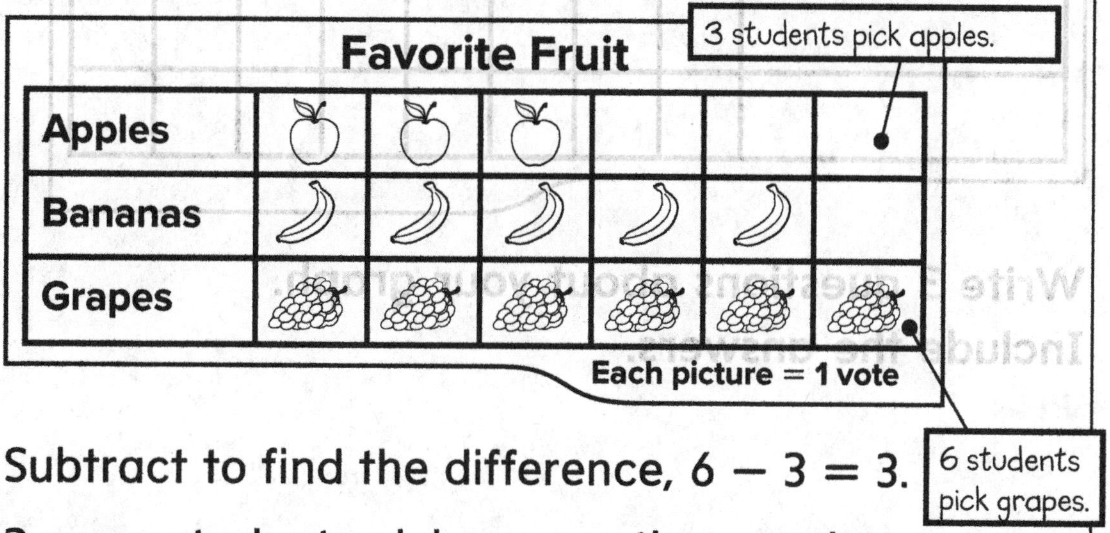

Subtract to find the difference, $6 - 3 = 3$.

3 more students pick grapes than apples.

Use the picture graph to answer the questions.

1. Which fruit did the most students choose?

____

2. Which fruit did the fewest students choose?

____

3. How many fewer students chose apples than bananas? ______ students

Solve Problems Involving Data

Name ____________________

Make your own picture graph about favorite carnival activities.

Write 3 questions about your graph. Include the answers.

Understand Equal Shares

Name ______________________________

Review

You can trace and fold shapes to see if they have equal shares.

A **whole** has **equal shares** when the parts are the same size.

Equal Shares	*Not* Equal Shares
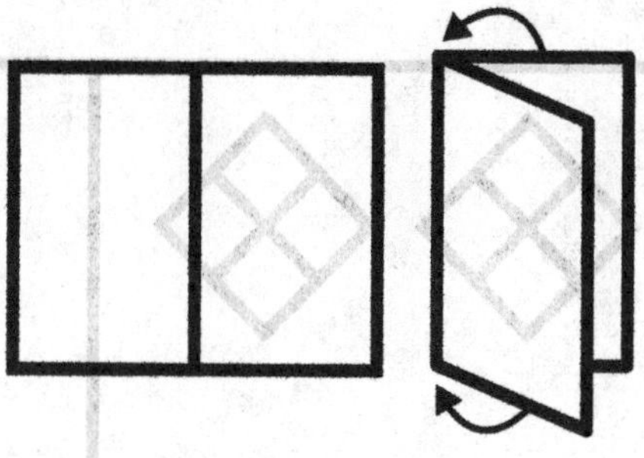	

Circle the picture that shows the original shape with equal shares.

1.

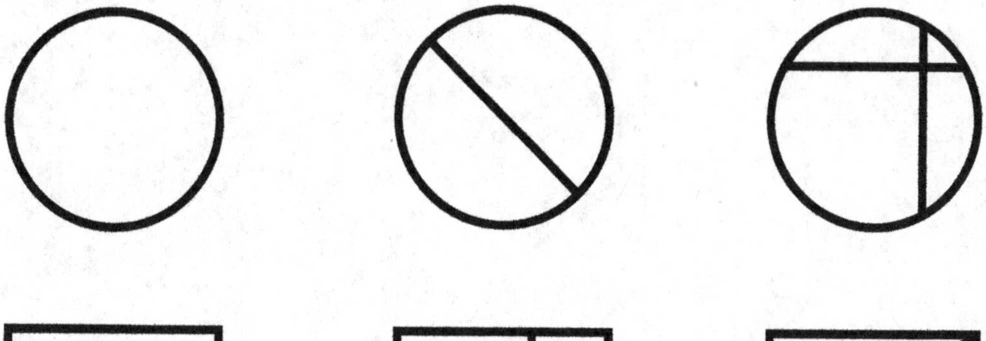

2.

Understand Equal Shares

Name ______________________________

Circle the parts of the train that have equal shares. Put an X on the parts of the train that do not have equal shares.

Partition Shapes into Halves

Name ______________________________

Review

You can use *half of* and *halves* to describe 2 equal shares. You can make halves in different ways.

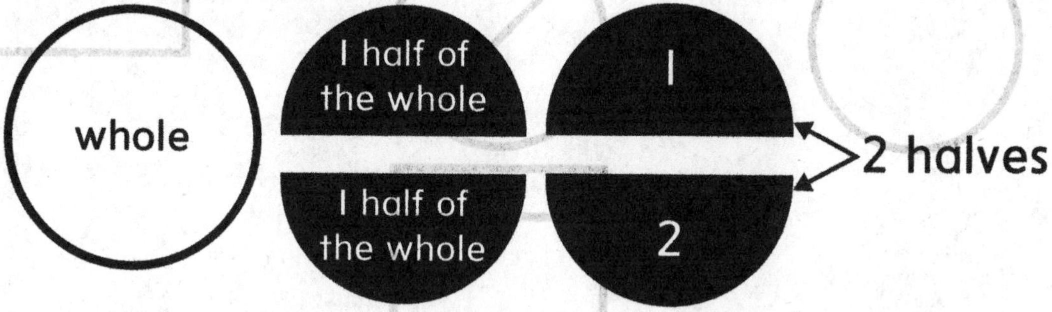

1. Which shapes show halves? Circle the shapes.

How can you make halves? Draw to show halves.

2.

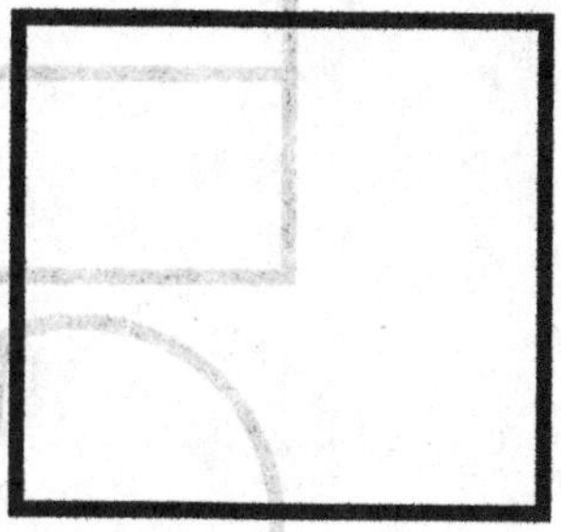

3.

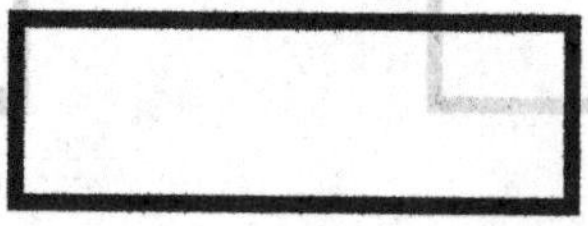

Partition Shapes into Halves

Name ______________________________

Match the halves with the whole. Find all the matches. Put an X over partitions that do not make halves.

1.

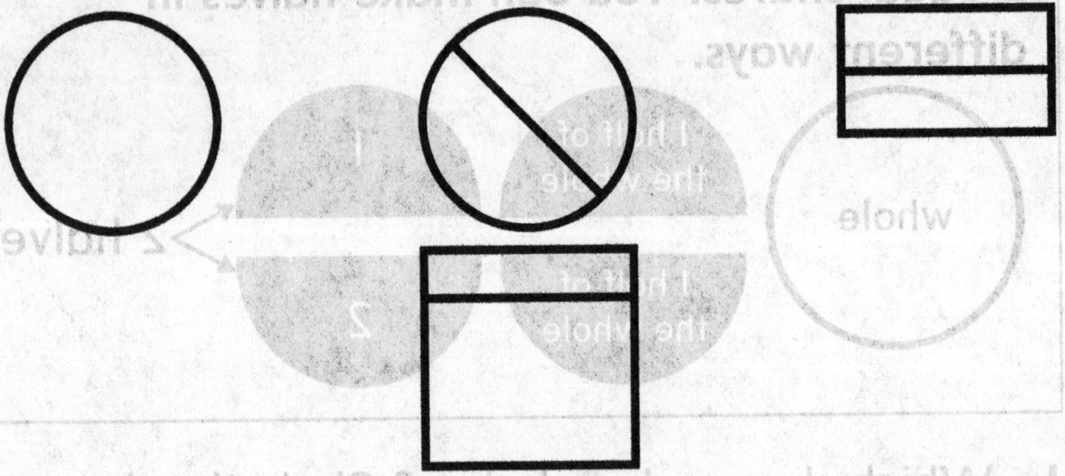

2.

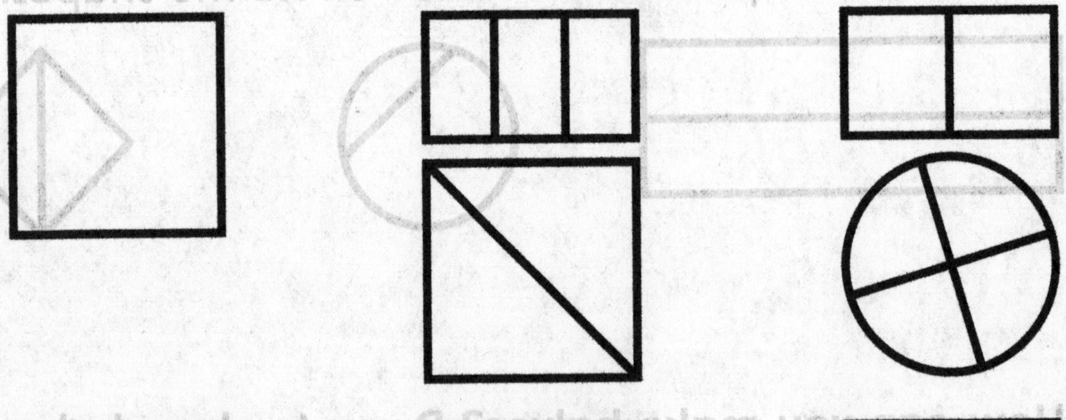

3.

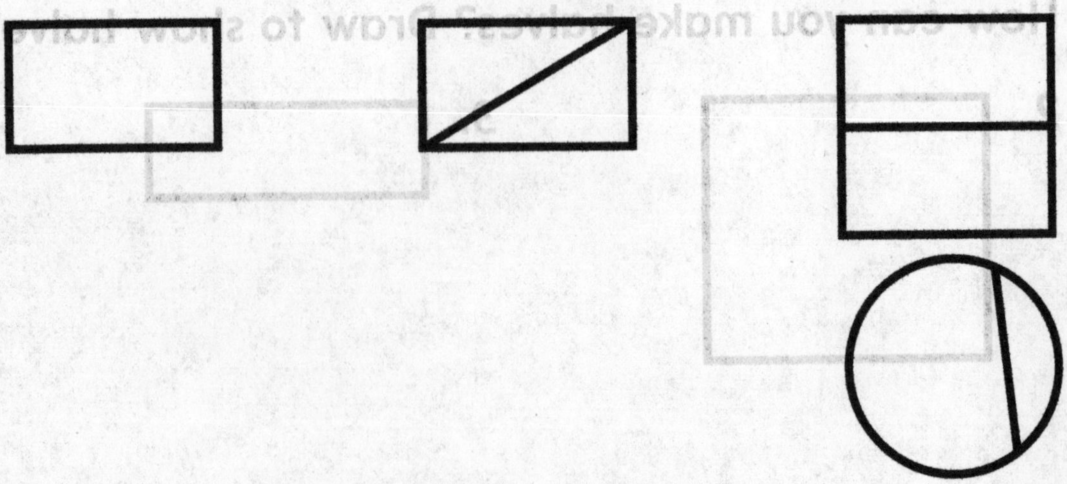

Partition Shapes into Fourths

Name ______________________________

Review

You can use *fourth of* and *quarter of* to describe 4 equal shares. You can make fourths in different ways.

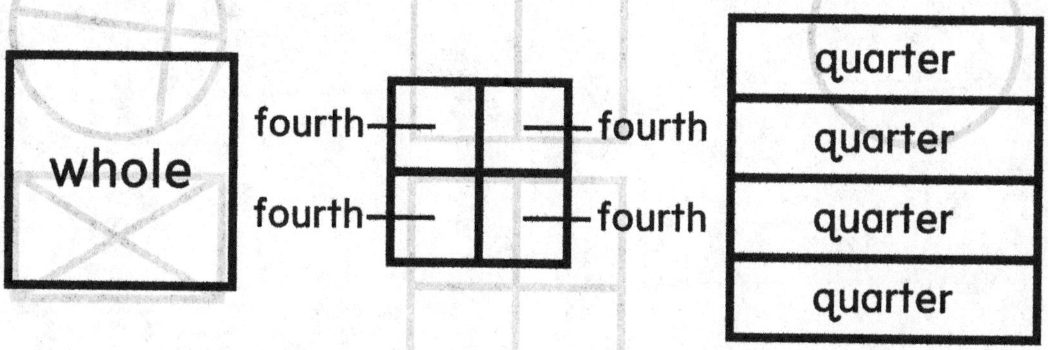

1. Which shapes show quarters? Circle the shapes.

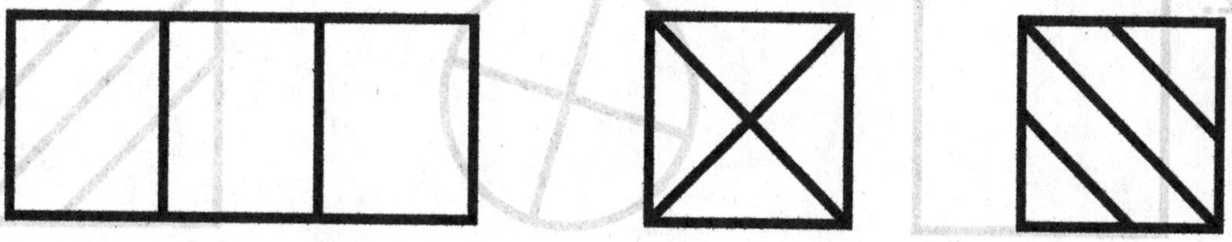

How can you make fourths?
Draw to show fourths.

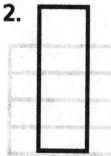

2.

3.

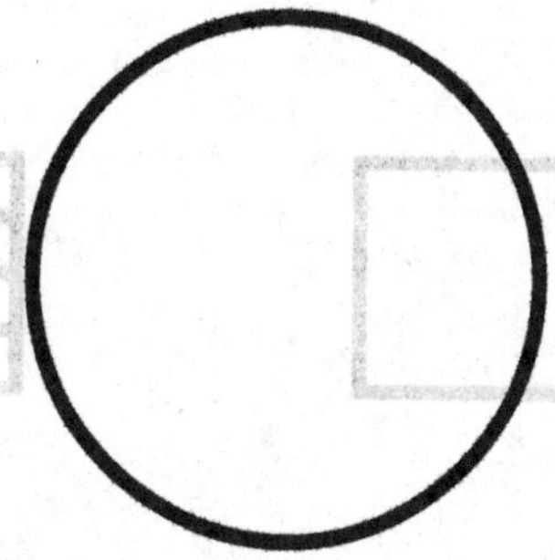

Partition Shapes into Fourths

Name ______________________________

Match the fourths with the whole. Find all the matches. Put an X over partitions that do not make fourths.

1.

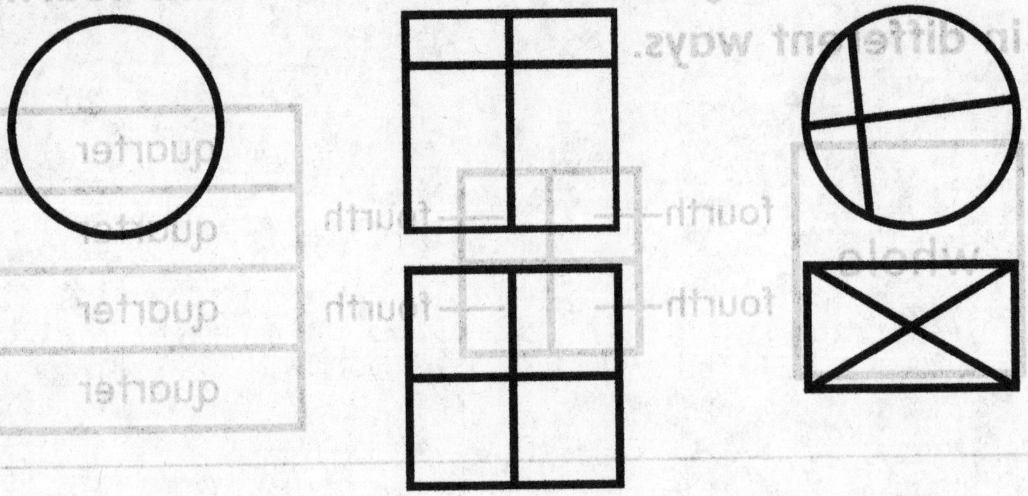

2.

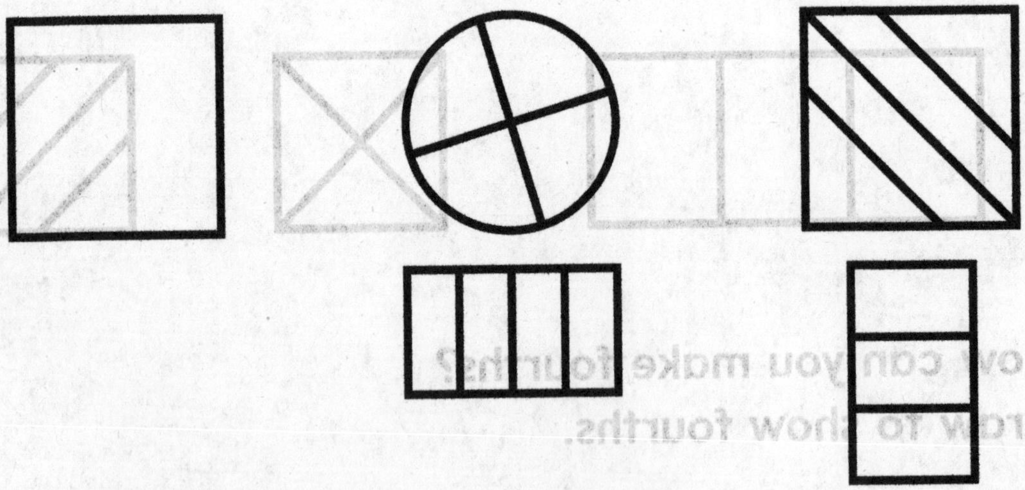

3.

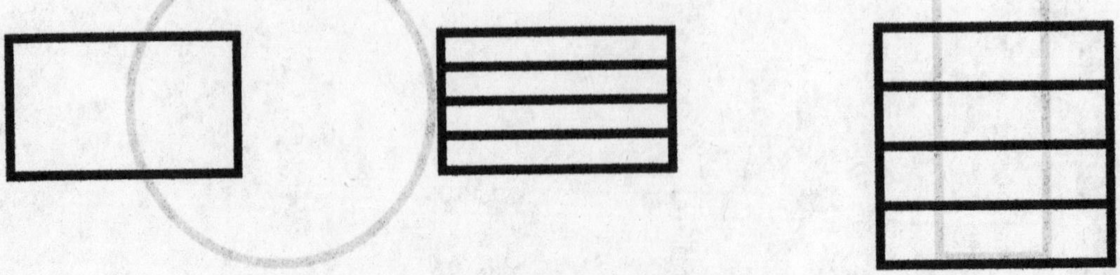

Describe the Whole

Name ______________________________

Review

You can describe a whole by telling how many equal shares.

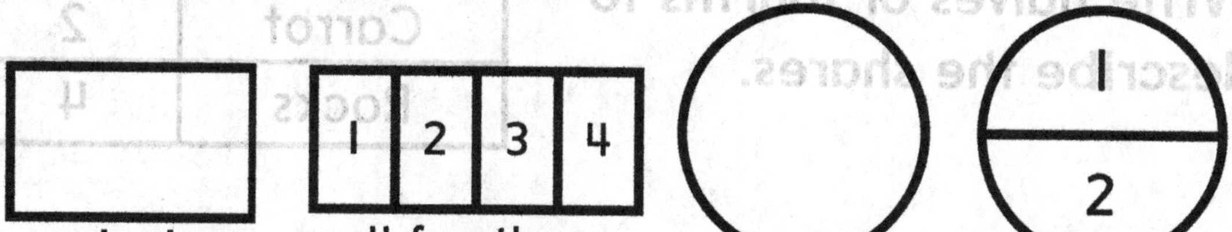

whole | 4 fourths 4 quarters | whole | 2 halves

1. Circle the shape that has 2 halves.

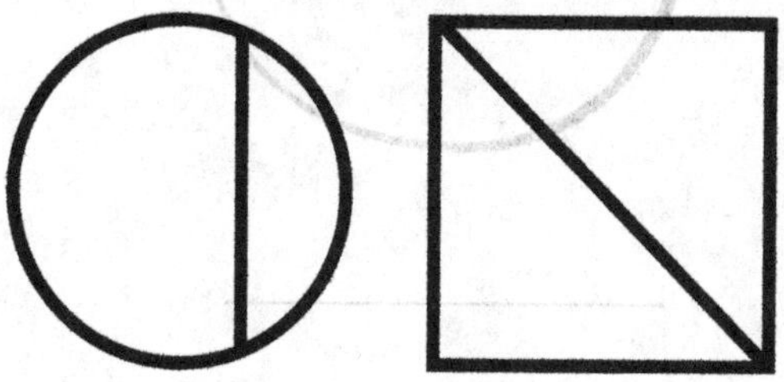

2. Circle the shape that has 4 fourths.

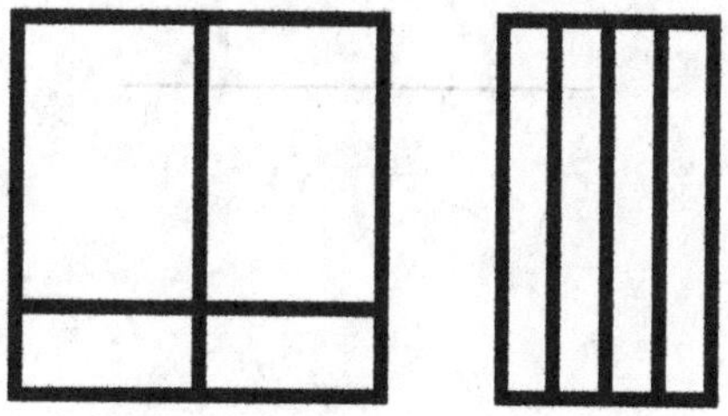

How many equal shares are there? Write the number.

3.

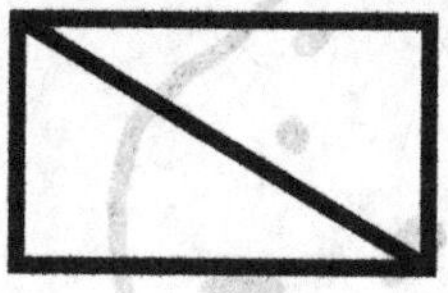

______

4.

______

Describe the Whole

Name ____

Ellen and her cousin are dividing some items. Draw to make equal shares. Write halves or fourths to describe the shares.

Items	Shares
Pillow	4
Orange	2
Carrot	2
Rocks	4

1. carrot

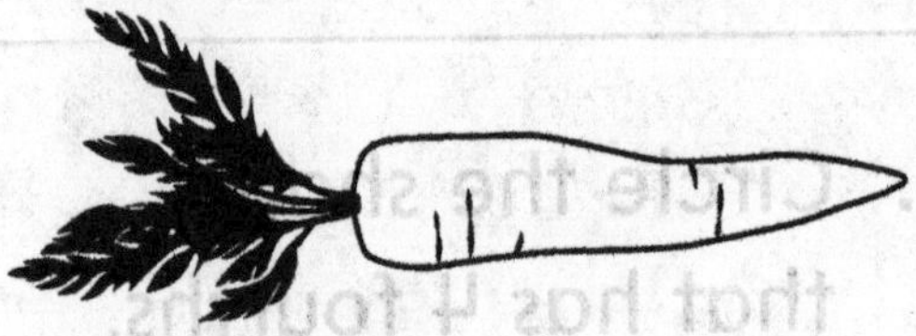

____

2. orange

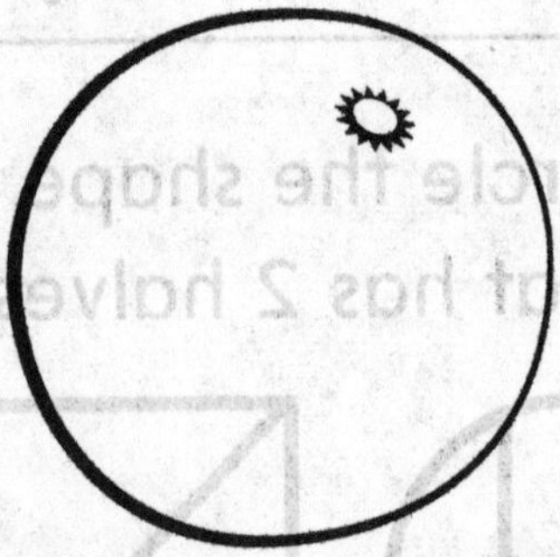

____

3. pillow

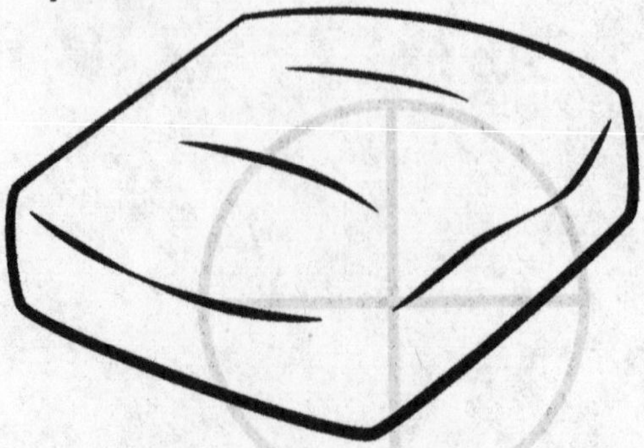

____

4. rock

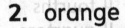

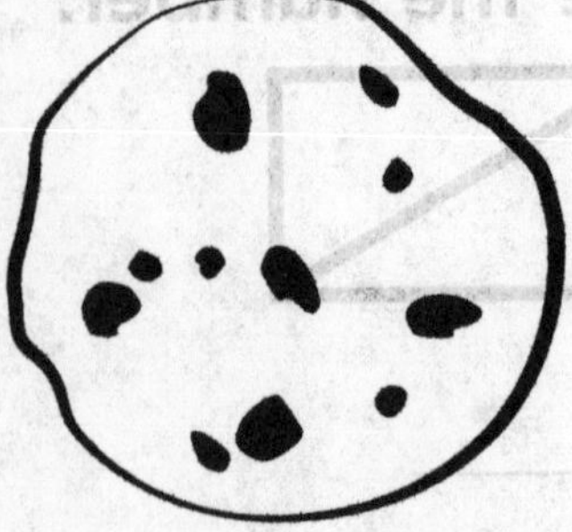

____

Describe Halves and Fourths of Shapes

Name ______________________________

Review

The same whole can have 2 halves or 4 fourths.

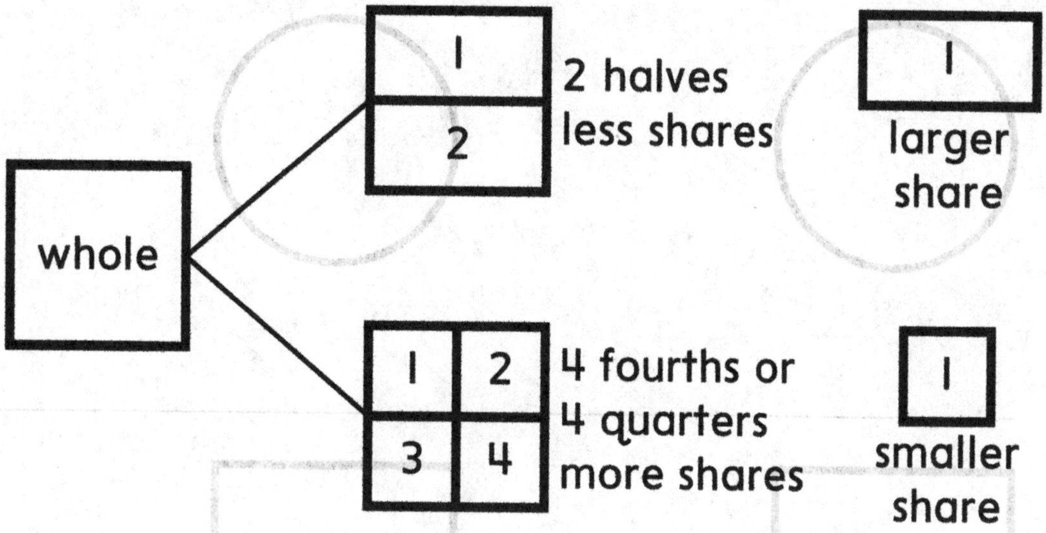

1. Which shows fewer equal shares? Circle the shape.

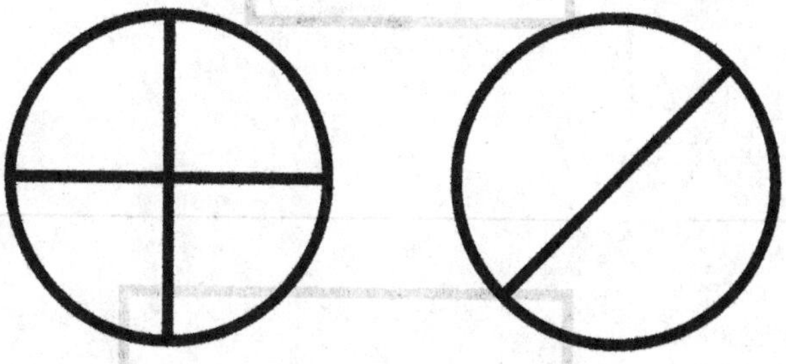

2. Which shows larger equal shares? Circle the shape.

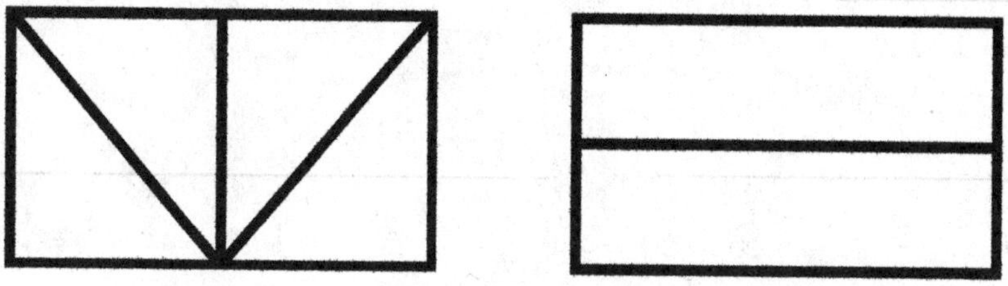

Describe Halves and Fourths of Shapes

Name ________________________________

Draw one way each shape can have 2 halves and 4 fourths. Write fewer, more, smaller, and larger under the correct drawing.

1.

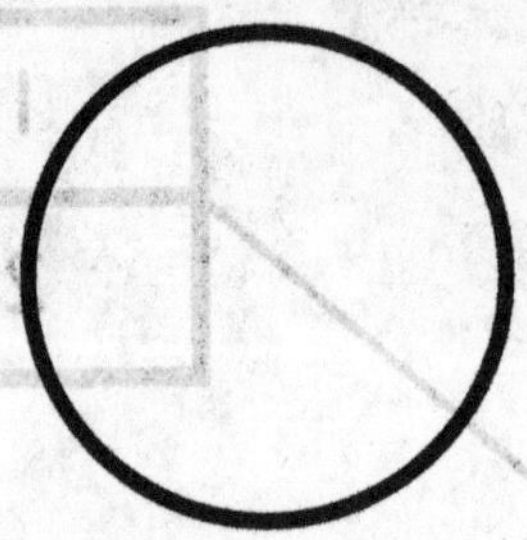

2.

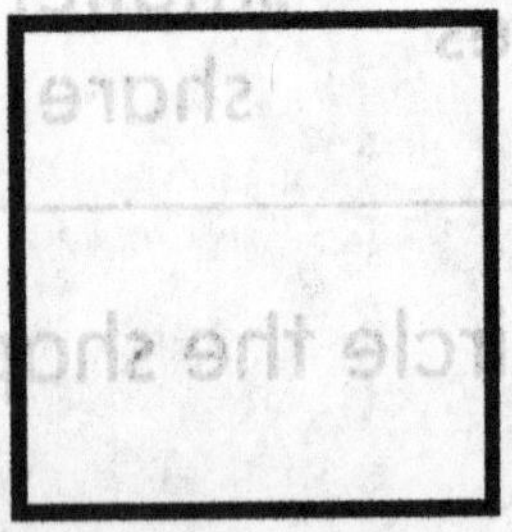

3.